TIME AND REALITY IN T
THOUGHT OF THE MAYA

by Miguel León-Portilla

foreword by J. Eric S. Thompson

translated by Charles L. Boilès and Fernando Horcasitas

We all live in time, but few of us realize that our time sense is a window on the world, and culture-bound.

Time and Reality in the Thought of the Maya is Miguel León-Portilla's exploration of one culture's remarkable concepts of time. For Evon Z. Vogt, Curator of Middle American Ethnology at Harvard University, it is a "work that will become a classic in anthropology."

Like no other people in history, the ancient Maya of Central America were obsessed with the study of time. Their sages framed its cycles with tireless exactitude. Yet the Maya preoccupation with time was not limited to mathematical calendrics; it has been a key – though little understood – trait in the evolution of their culture.

In this fascinating study, León-Portilla probes the question: What did time really mean for the ancient Maya – in terms of their mythology, religious thought, world view, and everyday life? In his analysis of various Maya texts and computations, he reveals one of the most elaborate efforts of the human mind to penetrate the secrets of existence.

For the ancient Maya, he concludes, time was the primordial reality – a divine cosmic atmosphere from which space and all living things derived their meaning.

TIME AND REALITY IN THE THOUGHT OF THE MAYA

In its study, they found norms for their everyday life, for their social, economic, and political systems. But above all, their passion for time was a concern for salvation, an attempt to find their place on earth and to spy on the mysteries of the universe.

The degree to which these concepts persist in the contemporary Maya world is described in a brilliant appendix by Alfonso Villa Rojas.

Miguel León-Portilla is director of the Institute of Historical Research and professor of ancient history of Mexico at the National University of Mexico, and is well-known for his work on Nahuatl culture, which includes *The Broken Spears*, a Beacon paperback.

TIME AND REALITY IN THE THOUGHT OF THE MAYA

MIGUEL LEÓN-PORTILLA

Foreword by J. Eric S. Thompson
Appendix by Alfonso Villa Rojas
Translated by Charles L. Boilès
and Fernando Horcasitas

Beacon Press *Boston*

Time and Reality in the Thought of the Maya was first published under the title *Tiempo y Realidad en el Pensamiento Maya: Ensayo de Acercamiento* by Universidad Nacional Autónoma de México.

Beacon Press books are published under the auspices
of the Unitarian Universalist Association

Simultaneous publication in Canada by Saunders of Toronto, Ltd.

Printed in the United States of America

9 8 7 6 5 4 3 2 1

Library of Congress Cataloging in Publication Data

León-Portilla, Miguel.
Time and reality in the thought of the Maya.
Translation of Tiempo y realidad en el pensamiento maya.
Bibliography: pp. 160–172.
1. Mayas—Philosophy. 2. Calendar, Maya.
I. Villa Rojas, Alfonso. II. Title.
F1435.3.C14L413 970.3 72–6229
ISBN 0–8070–4664–7

CONTENTS

In memory of Dr. Manuel Gamio (*1883–1960*)

FOREWORD

I much appreciate the honor of being asked by you to write an introduction to your study of the Maya concept of time. I have tried to comply with your request, but after several failures I found it easier to write you this open letter. It is partly congratulatory, but partly, as you will see, a sort of *apologia pro vita mea,* at least for that part of my life involved with study of the Maya philosophy of time.

First, let me congratulate you for having expanded your interests to encompass that very specialized field of Maya time. Courageously, you are battling against one of the greatest defects of modern research, namely, the trend to know more and more about less and less. As a victim of that process, I feel somewhat like a backsliding sinner congratulating a saint on overcoming temptation. Forty years ago, in my first job at Field Museum, Chicago, I had charge of everything archaeological and ethnographical from the Rio Grande to Tierra del Fuego; ever since I seem to have complacently watched the frontiers of my knowledge contract. Alas! the shrillness of the cock's crow is not related to the smallness of his dunghill. In reversing that modern trend you set others an example I hope they will follow.

Let me also offer my heartiest congratulations on your very penetrating study of the Maya philosophy of time; it is full of fresh insights. You have looked at this old problem from a new angle of vision, from the clearer atmosphere of the heights which nourished Nahuatl culture.

Perhaps it is the pessimism of many years of playing hide-and-seek with Maya reality which bids me whisper to both of us a word of caution, which I hope your ears are sharp enough, and mine near enough, to hear. Both your and my attempts to study the Maya philosophy of time are, I fear, likely to fall far short of complete success.

It is, perhaps, as irrational to expect a satisfactory penetration of the mystic and emotional aura of the Maya philosophy of time by a creature of twentieth-century Western culture as it is to hope for a balanced, sympathetic and understanding study of the ecstasy of St. Francis from the pen of a militant atheist of our materialistic age. Our outlooks are too far from those of the Maya and, on top of that terrible handicap, there are so many aspects of the problem which are imperfectly known or completely unknown to us. The atheist student of St. Francis has at his disposal incomparably richer sources than we can ever dream of having.

Many years ago, with unjustified arrogance I compared my ambivalent position as a Maya student—in the picture, but not of it—to that of the humble donor whose portrait is allowed to appear in the corner of some great religious painting of the early renaissance he has commissioned. I had meant to convey that, at best, the student, like the donor, is a nonparticipant, but is honored by being allowed an imaginary participation from afar in the proceedings, but I was ranking myself too high; neither you nor I will ever have the insight into Maya mysteries that the kneeling donor had in that age of faith. I fear we shall never attain a corner of the canvas.

Perhaps I have been unduly pessimistic in assessing our problems and your ability to meet them. My emotional insight into Maya mysteries will always be from the far side of that deep chasm which divides Maya culture from ours; I can never hope to be a participant as was that kneeling donor who was both outside that particular scene of mystery or miracle, but at the same time an actor in it. You perhaps may not have to remain on that far brink; your

Nahuatl studies may show you some goat track down the chasm and up the other side.

The *Ah Beeob* were obscure Maya deities or, more probably, obscure aspects of well-known gods. Their name means "They of the road," and their duty was to clear the paths for travelers. May they sweep a clear path for your exploratory footsteps.

J. Eric S. Thompson

January 1967

INTRODUCTION

The Maya of Central America were masters in the art of measuring time since their days of Classic splendor—centuries 2 to 9 A.D. Today, the extraordinary precision of their calendrical systems and a good deal about their chronology and astronomical science can be appraised thanks to the research of not a few scholars. The list would begin with Fray Diego de Landa's *Relación* in the sixteenth century and, in modern times, with the studies of Förstemann, Goodman, Seler, Spinden, Martínez Hernández, Beyer, Teeple, Morley, Berlin, Barthel, Satterthwaite, Lizardi Ramos, and Thompson.

These studies, and others regarding different aspects of Maya culture, have permitted to attempt here a new kind of approach: to study the theme of time in its direct relationship with the world view of these ancient Americans. This book will not augment what is known about the calendrical computations or the astrological predictions of their priests and sages. In a word, the question will be: What did time really mean for the Maya in their mythology, religious thought, world view—and everyday life?

In recent decades, the British scholar J. Eric S. Thompson has, better than anyone else, penetrated the problem, analyzing the complex of ideas that comprise what he calls the "Maya philosophy of time." He comments that "one should point out one important element of Maya civilization, the overwhelming preoccupation with time. Indeed, one might say the Maya philosophy of time" (1952:37). And when treating of this in his work *The Rise and Fall of*

Maya Civilization, he draws attention to the need to probe the meaning the ancient sages gave to time in order to achieve a more adequate understanding of their culture.

> No other people in history has taken such an absorbing interest in time as did the Maya, and no other culture has ever developed a philosophy embracing such an unusual subject.
>
> THOMPSON 1954:162[1]

It is precisely the aim of the present work to continue the line of investigation initiated by Thompson. In attempting this, account has been taken of his findings and those of other Mayists, as well as of our own research on the basis of the inscriptions, the codices and native texts reflecting something of ancient Maya thought concerning time.[2]

As an introductory framework for this study, it will be necessary to recall a few well-proved and impressive facts in respect to the chronological and astronomical knowledge of the Maya. This will help give insight into the proportions of what was for them an insistent preoccupation. The following step—the analysis of the complex of symbols, of the glyphs with temporal connotation that have been deciphered, and of the late texts of mythological, religious,

1. See also Thompson's Appendix 4, "Maya calculations far into the past and into the future," in *Maya Hieroglyphic Writing,* New Edition, Norman: University of Oklahoma Press, 1960, 314–316.

2. Apart from Thompson's *Maya Hieroglyphic Writing,* indispensable instruments for the study of glyphs and inscriptions having temporal connotation or diverse relationships with the Maya mythological and religious symbolism are the following two fundamental works: Zimmerman, Günter, *Die Hieroglyphen der Maya-Handschriften,* Universitat Hamburg, Abhandlungen aus dem Gebiet der Auslandkunde, vol. LXVII, Reihe B, Hamburg, 1956; Thompson, J. Eric S., *A Catalog of Maya Hieroglyphs,* Norman: University of Oklahoma Press, 1962.

and calendrical content—is directed toward the search for the essence and meaning of this concern, so directly related to the core of the Maya world view throughout its cultural evolution.

Most certainly the task will encounter not a few problems as yet unsolved. Eric Thompson, to whom deep appreciation is expressed for having read this work and given some valuable suggestions, reiterated how difficult it is "to expect a satisfactory penetration of the mystic and emotional aura of the Maya philosophy of time by a creature of twentieth-century Western culture." But he who so sharply points at the root of problems inherent in all processes of historical and anthropological comprehension, also speaks of the possibility of finding "some goat track down the chasm and up the other side." And as a matter of fact, in his own works he confirms that with an awareness of one's inescapable limitations, it is not futile to attempt new forms of research.

From the beginning, it must be recognized that in the present state of investigations, it is impossible to establish adequately a scheme of the evolution of the mythological and religious thought of the various groups embracing the great cultural tree of the Maya. It would be dangerously naïve to presuppose that world view and religious concepts were one and the same in different areas and diverse periods covering what is generically known as "Maya culture." Thus the mere theme of its evolution immediately poses a mass of problems.

Nonetheless, the recognition of these and other adverse circumstances does not invalidate the existence of a rich series of indigenous sources, possibly a path to some findings about the different Maya forms of meaning concerning the subject of time. Eventually, an approach toward such an "overwhelming preoccupation," with no break in Maya history, could cast light on the problem itself of variants and affinities in the different stages and areas in which this culture flourished. As will be seen, in spite of the differences in styles, symbolism, and other patterns and institutions of the various Maya groups, their interest in time never disappeared. This affirmation is valid at least from the moment the first stelae of the Classic period were erected, about the third century A.D., until the appearance of the Postclassical codices and of the more recent colonial manuscripts of the Yucatec Maya. In these manuscripts are preserved the prophecies of the *katuns* or twenty-year cycles and many other calendar texts. Such is the case of the books of *Chilam Balam* from Yucatán and other native productions from Chiapas and Guatemala.[3] Moreover, it can be demonstrated that quite a few traits, essential to the pre-Hispanic concept of time and space, have survived among contemporary groups belonging to the same family. This will be shown in the Appendix—an ethnohistorical study by the Mayist scholar Alfonso Villa Rojas.

As emphasized by Evon Z. Vogt (1964:35), time being a "key trait or culture pattern"[4] in the evolution of the Maya, its study will help penetrate more deeply some of the most imbedded symbolism and forms of conceptualization in

3. More recent proof of this continued interest is offered by the discovery of texts of similar content and of calendars based on Maya tradition, presently being used among groups in very distinct areas. To cite a single example, in 1936, Alfonso Villa Rojas discovered the *Chilam Balam de Tusik* (see Villa Rojas, Alfonso, *The Maya of East Central Quintana Roo,* Washington, D.C., The Carnegie Institution of Washington, Publication 559, 1945.)

4. Vogt, Evon Z., "The Genetic Model and Maya Cultural Development" in *Desarrollo cultural de los mayas,* edited by E. Z. Vogt and A. Ruz, Mexico, Seminario de Cultura Maya, Universidad Nacional, 1964, p. 35. As related to the matter here to be discussed he points out the need for a deeper study "not only about the technical aspects of the intellectual achievements that expressed this preoccupation with the measurement of time, but also about its probable origins and deep meaning in the life of the Maya."

what was perhaps the original nucleus of this Middle American world view. This is even more justified if it is considered that in the diverse stages of Maya cultural sequence mythological and religious thought were always in close relation with calendrical and chronological preoccupations. Actually, Eduard Seler and more recently David H. Kelley, writing about the first appearance or "birth" of some Maya gods,[5] give good examples of the constant relationships between deities and calendar symbols present on monuments of the Classic period and also in the later Maya codices.

When inquiring into the significance of the Maya concept of time in terms of their world view, differences in symbolism and thought throughout their evolution will be pointed out. Nevertheless, elements and traits that perhaps constituted a kind of common substratum in the spiritual atmosphere of these peoples also will be emphasized. In an enterprise such as this all precautions and critical awareness will not be in excess. At least what up to the present has been investigated constitutes a highly attractive invitation. A preliminary résumé of Maya chronological and astronomical achievements will help in gaining a more thorough orientation toward the subject under consideration.

In spite of the limitations imposed by the sources and by the radical sum of differences with our mentality, from the point of view of the history of ideas, the thought of the Maya offers the possibility of an approach to original and extraordinary concepts concerning an age-old theme: time. In many forms time has captured the attention of man, who is essentially a temporal being, because, living in time, he is

5. Kelley, David H., "The Birth of the Gods of Palenque," in *Estudios de Cultura Maya,* vol. V, pp. 93–134. See also Seler, "Maya Handschriften und Maya Götter" in *Gesammelte Abhandlungen,* vol. I, pp. 357–366, and "Ueber die Namen der in der Dresdener Handschrift angebildeten Maya-Götter," *ibid.,* pp. 367–396.

conscious of it and in exceptional moments discovers more than one mystery in it.

I attest, once more, my gratefulness to J. Eric S. Thompson, Alfonso Villa Rojas, and Demetrio Sodi, who read and commented on the manuscript of the present work, and to Víctor Manuel Castillo Farreras, who reproduced the glyphs and other illustrations.

Miguel León-Portilla

15 April 1967
Instituto de Investigaciones Históricas
Universidad Nacional
Mexico City

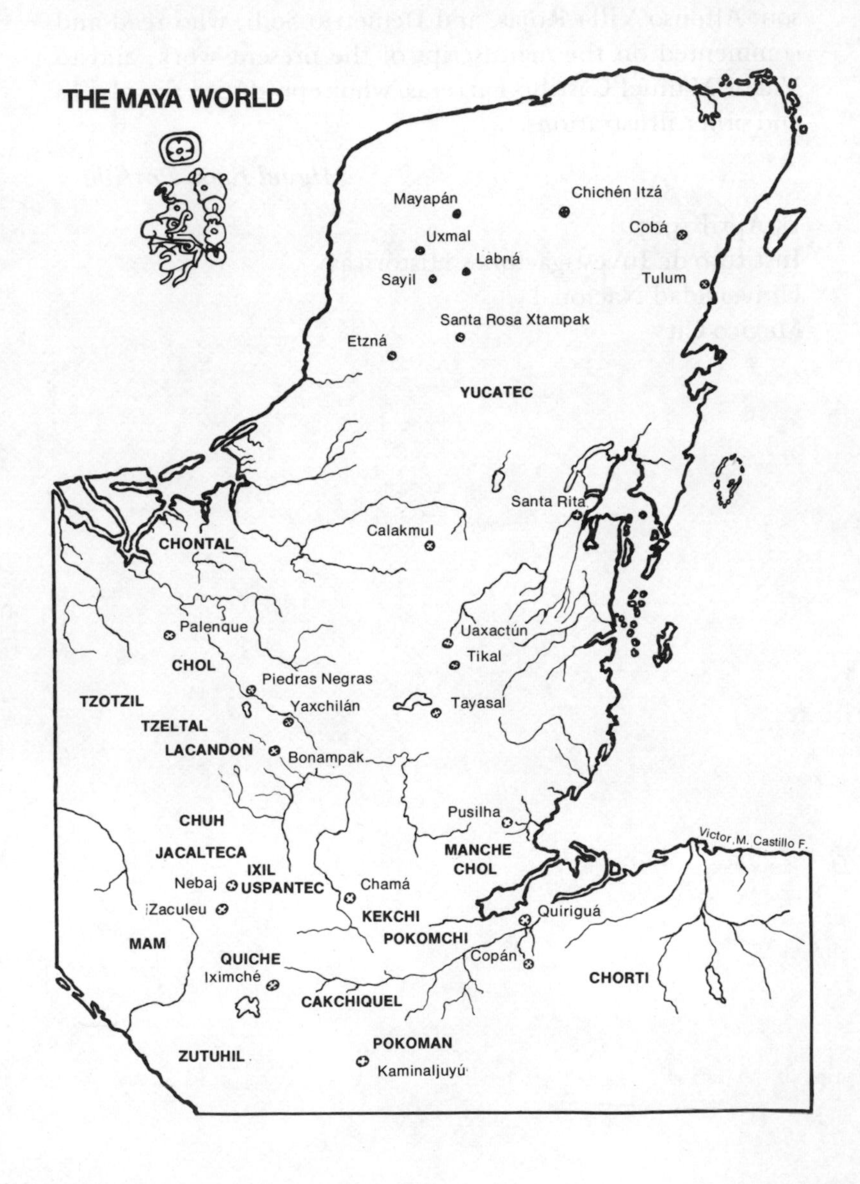
THE MAYA WORLD
Mayapán
Chichén Itzá
Cobá
Uxmal
Labná
Sayil
Tulum
Santa Rosa Xtampak
Etzná
YUCATEC
Santa Rita
Calakmul
CHONTAL
Palenque
Uaxactún
Tikal
CHOL
Piedras Negras
TZOTZIL
Yaxchilán
Tayasal
TZELTAL
LACANDON
Bonampak
Pusilha
CHUH
Victor M. Castillo F.
JACALTECA
MANCHE
CHOL
IXIL
Nebaj
USPANTEC
Chamá
Zaculeu
KEKCHI
Quiriguá
POKOMCHI
MAM
QUICHE
Copán
Iximché
CHORTI
CAKCHIQUEL
POKOMAN
ZUTUHIL
Kaminaljuyú

CHAPTER I

THE MAYA CONCERN WITH CHRONOLOGY

A retracing of the main chronological pursuits of the Maya is a step required in trying to understand their concept of time. Among their sages, outstanding features of mathematical knowledge were applied to calendar computations and endeavors in astronomy. Restricted to arithmetic and geometry, Maya mathematics from the beginning of the Classic period (around the third century A.D.) included, nonetheless, two extraordinary and closely related discoveries: the concept of zero, principally as a symbol of completeness, and a vigesimal counting system in which unities acquired value according to positional functions (Figure 1).[1] What is known about this today comes from the inscriptions in the Maya codices and on stelae erected in the early centuries of the Christian era.[2] There is no parallel in the Old World until around the eighth century A.D., at which time Hindustani scholars arrived at a concept of zero within a decimal system of numeration. Europe was not to possess these discoveries until many centuries later, benefiting from

1. Concerning the diverse studies and controversies about the meaning of the "zero" concept and glyph among the Mayas, see the analysis by César Lizardi Ramos, "El cero maya y su función," in *Estudios de Cultura Maya,* vol. II, Seminario de Cultura Maya, Universidad Nacional de México, 1962, pp. 434–453.

2. Among the most ancient inscriptions attesting to this should be

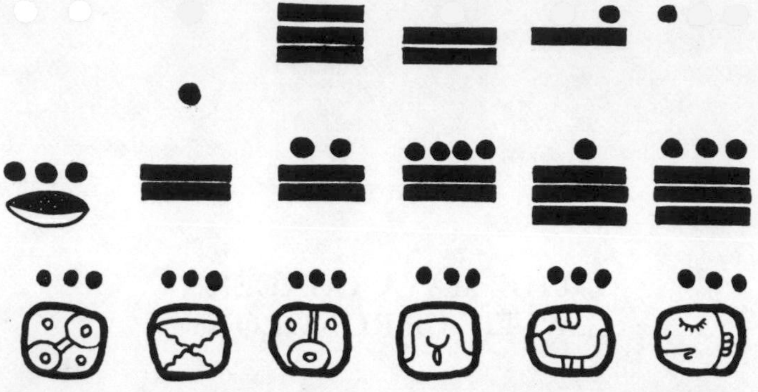

Figure 1. An example of the use of numerals on a multiplication table and divinatory almanac in the *Dresden Codex* (44-b). Beginning above, on the extreme right, reading from top to bottom, the following ciphers and glyphs appear: 3. 18. 3 *Cimi;* 7. 16. 3 *Kan;* 11. 14. 3 *Ik;* 15. 12. 3 *Ahau;* 1. 1. 10. 3 *Etznab;* 2. 3. 0. 3 *Lamat*

diffusion by the Moslem civilization through Spain.

Having mastered these findings, the Maya developed temporal computations of high precision. Among those to be recalled here are the various calculations with reference to the solar year, to what is now called the synodic revolutions

---

mentioned the Leyden plaque (giving a date corresponding to A.D. 320), Stela 9 of Uaxactun (A.D. 328), and the more recently discovered Stela 29 of Tikal (A.D. 292).

From an epoch immediately prior to the Classic period, in regions near the Maya area, come other calendar inscriptions in which the numerical value assigned to numbers according to their position seems to have already been adopted: Stela C of Tres Zapotes (Veracruz), probably dating from 31 B.C., the Tuxtla statuette (A.D. 162), as well as the stela of El Baúl (Guatemala)—among others found in the cultural area of Izapa (Chiapas)—dating from 256 years prior to the inscription on Stela 29 of Tikal.

These testimonies, similar to the calendar inscriptions on the Dancers stelae of the ancient Monte Albán I period in Oaxaca, seem to indicate the appearance and early diffusion in Middle America of a growing interest in measuring time, the art and science to be developed to the greatest extent by the priests and sages of the Mayan lowlands.

of Venus, and to the lunation periods, including the tables elaborated for predicting eclipses. Mathematical knowledge also made it possible to register any date in their so-called Long Count or Initial Series system and, of greater importance, the corresponding correction formulae for adjusting and correlating with distinct astronomical cycles the dates expressed in the function of their calendar.

Within the field of chronology itself, the Maya made various, equally extraordinary accomplishments. As noted by Thompson (1960:149), their sages apparently conceived of time "as something without beginning or end, and therefore one could project one's calculations farther and farther into the past without ever reaching a starting point." Among other examples are two especially impressive computations (Figure 2):

Figure 2. Computations of remote dates in the past, after Thompson. *a*) From Stela F at Quiriguá, bearing the date 1 *Ahau,* 13 *Yaxkin:* 91,683,930 years into the past. *b*) From Stela D, also at Quiriguá. It bears the date 7 *Ahau,* 3 *Pop:* some 400,000,000 years before the erection of this monument

> On one stela at the city of Quiriguá accurate computation sweeps back over ninety million years; on another stela at the same site the date reached is some four hundred million years ago. These are actual computations stating correctly day and month positions. . . .
>
> THOMPSON 1956:23

But along with this unique concept of time without limits either in the past or future,[3] the Maya established a reference point, marking a sort of beginning for their chronological era. Accordingly, all calendar inscriptions on their stelae are computed with reference to this beginning point which, in terms of our own calendar, is situated at 3133 B.C.[4] Various investigators, including Morley (1947:282–284), considered this reference point a fixed base in Maya chronological computations. But rather than restrict their concept of time without limits, this date seemed to refer to an especially significant event in their past. This, as Thompson (1960:149) has indicated, could supposedly be "regarded as the last creation of the world," i.e., the "age" and "sun" in which men were made from corn, according to the narrative given in the *Popol Vuh* (Recinos 1953:174–176). Surprisingly, Maya chronology of the Classic epoch thus developed with a belief in limitless time yet with a reference point in order to effect computations.

To record the passage of time, priests and sages began to erect stelae and to register dates especially significant for them. Beginning with Stela 29 of Tikal, Guatemala, on which is posted the oldest known Maya calendar inscription, dated A.D. 292, the erection of these monuments reaches an extraordinary diffusion. In the lapse between the aforementioned date and A.D. 928, a date appearing on a rough stela found at San Lorenzo near La Muñeca in the Mexican State of Campeche, the art and science of inscriptions materialized in a multitude of centers in the vast Maya zone.

The system known as the Long Count or Initial Series achieved an exceptional diffusion in almost the whole exten-

3. Regarding other Maya computations concerning remote dates in the future, see Thompson 1960:314–316.

4. Numerous examples can be cited of solar anniversaries computed on the basis of the original beginning point, 4-*Ahau* 8-*Cumkú* or the year 3113 B.C. Among them are those recorded on stelae such as No. 8 of Piedras Negras, No. 9 of Calakmul, and No. 8 of Naranjo.

sion of the lowland Maya area. With clarity diverse computations on monuments are expressed, in terms of the Long Count, by means of notations in which zero and positionally functioning unities of the vigesimal system register distinct cycles or periods of time. In the uppermost part of a Long Count inscription appears a large hieroglyph, the so-called introductory glyph of the series of computations. This hieroglyph is always formed of a *tun* (year) sign, accompanied by its corresponding affixes, and of a variable central element which is the glyph of the deity presiding over the month, or twenty-day cycle, in which the computed date occurs. Immediately following this, always reading from left to right and continuing from top to bottom, appear the various unities in the following order: first, the *baktuns,* cycles of 360 × 20 × 20 days = 144,000; then, the *katuns* (360 × 20 days = 7,200); the *tuns* (360 days), the *uinals* (20 days); and, finally, the *kins* or days.

Almost at a glance one may thus compute the periods arrayed in a specific inscription. At the same time, these and many other cycles known to the Maya—the *tuns* (360-day years), the approximate years of 365 days (occasionally called *haab*), the lunations and synodic revolutions of the planets—were coordinated with the typically Central American count of 260 days, known today in Yucatec as *tzolkin.* This coordination was achieved by using multiples common to said cycle and to other astronomical measurements.[5] The count of 260 days, used by all the high-culture peoples of ancient Mexico, had come to be a fundamental element in the Mayas' computations.

5. For an explanation of the nature of *haab* and *tzolkin,* as well as their interrelationships, see Morley 1947: 272–279; Thompson 1960: 66–128.

The equivalent calendar systems among the Nahua-speaking peoples are the *xiuhpohualli* and the *tonalpohualli* that have been studied by Alfonso Caso, among others, in "El Calendario Mexicano" (1958:41–96).

In finally coming to express in terms of a given month and day the periods computed in the inscriptions, it was possible to establish a date in which error was already suppressed by dint of the correction and adjustment formulae. The date so established resulted in correlation, not only with an annual cycle or with a fifty-two–year period (as in the case of the Aztec calendar), but with the entire Maya chronological system.

A mathematical analysis of this permits a more than surprising conclusion: the dates given as calendar expressions, according to the Long Count system, would never again be repeated until after a period of 374,440 years. This, as Morley (1947:289) affirms, is "a truly colossal achievement for any chronological system, whether ancient or modern."

A good illustration is offered by the classic example of Stela E of Quiriguá in Guatemala[6] (Figure 3). Reading from left to right and from top to bottom, one encounters the following calendar values:

The introductory hieroglyph of the series appearing with the sign of the year and the symbol of the god presiding over the corresponding twenty-day cycle, in this case that of *Cumkú*.

9 *baktuns*
(9 periods of 144,000 days)

17 *katuns*
(17 periods of 7,200 days)

0 *tuns*
(0 periods of 360 days)

0 *uinals*
(0 periods of 20 days)

0 *kins*
(0 periods of one day)

date 13-*Ahau*
(computed from the chronological point of departure)

Summing up the days arrayed in the inscription of this stela, the expressed computation reckons 1,418,000 days into the past.

6. Among others who discuss this classic example, see Morley 1947:337–338; Anders 1963:148–149.

Figure 3. Stela E from Quiriguá (western side), after Morley, showing the initial and supplementary series. The date in the long count is: 9 *baktuns,* 17 *katuns,* 0 *tuns,* 0 *uinals,* 0 *kins:* 13-*Ahau* 18-*Cumkú* (A.D. 771). This inscription appears here according to Morley's simplification. In reality the hieroglyph which represents the date 13-*Ahau* is actually situated on the stela in the line immediately below. The glyphs which, in this arrangement made by Morley, appear on the lower line, should occupy, united in a single block, the place of 13-*Ahau.* A reproduction of the stela appears in Maudsley, *Archaeology, Biologia Centrali-Americana,* vol. II, plate 32

The lower notations of this stela pertain to a grouping Maya scholars call the "supplementary series." Among other things there appear the lunar age, the lunar month position, and on the extreme lower right, the number and glyph of the month within the solar calendar, the 18-*Cumkú,* which must be correlated with the number and sign of the day, that is, 13-*Ahau*. Thus are expressed the 1,418,000 elapsed days equivalent to a date that occurs for the first time on the day 13-*Ahau* of the month 18-*Cumkú*.

The same Maya of the Classic era also often inscribed another manner of computation, the Secondary Series or "distance numbers," establishing by addition or subtraction the place corresponding within the Long Count to one or various dates that were not a termination of a specified period. Thus, apart from employing a new chronologically related formula, it was possible to express any date in rigorous function of its completeness. This was of fundamental importance to the Maya since their time measurements, as compared to ours, referred not to the beginning of a new cycle—day, month, year—but to its total reality already elapsed and completed.

The Classic Maya system of the Long Count became simplified in later centuries (from the eleventh on) when replaced by the so-called *u kahlay katunob* (count of the *katuns* or twenty-year periods), in which one single hieroglyph could express the day that concluded the corresponding period or *katun*.[7] This new system was to have great importance—along with the computations of the solar year and with the *tzolkin* or count of 260 days—as much for reg-

7. The date of the final day in each *katun* always coincided with an *Ahau* day, the sign of the sun, as can be seen in the celebrated "wheel of the *katuns*" preserved by Landa (1938:204) or in the series of *katuns* included in the *Paris Codex* (pp. 1–11) and in various books of the *Chilam Balam* that abound in recollections and predictions considered as attributes or "burdens" of the diverse *katuns.*

istering the principal events of the past as for providing the prophecies of priests and sages with a framework (Figure 4).

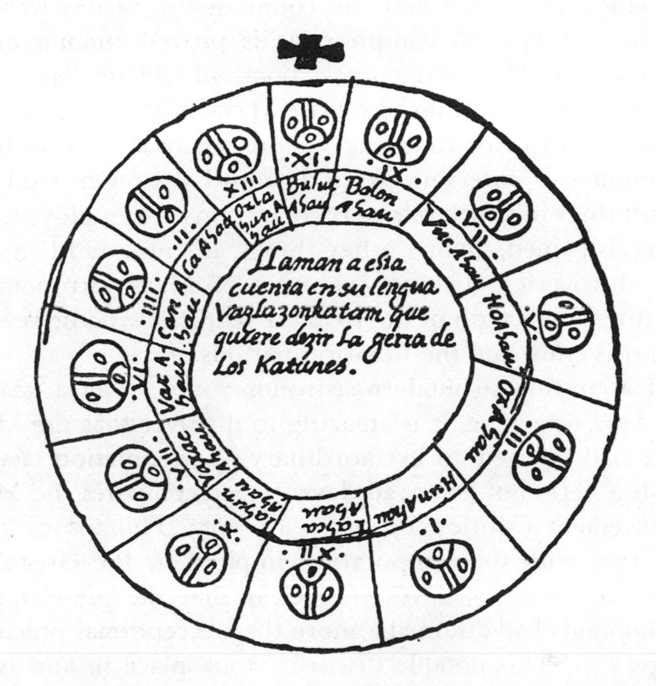

Figure 4. Wheel of the *katuns,* according to Landa. The passing of each *katun* should be read clockwise (20 X 360 days = 7,200). The *katuns* receive their calendrical designation depending on the name of the last day of the same, which is always an *Ahau,* accompanied by a numeral between 1 and 13. Only until after 256 years of 365 days, approximately, will the same date be repeated; that is to say, an *Ahau* day with the same number as the termination of a *katun*

The deepest meaning of what must be called the prophetic and astrological obsession of the Maya is a point to be investigated in this book. For the moment it can be said that precisely this kind of interest explains in part the development achieved by the measurements of time in this culture. This applies especially to the *tzolkin,* whose prophetic implications affected the entire people. This 260-day count directed the norms applicable to all important acts in life. Actually, the *tzolkin* and the count of the *katuns* were to outlive the Spanish conquest, as is proved, among other texts, by the celebrated Yucatec books of *Chilam Balam*.

The precision achieved by the Maya in their various systems for measuring time was obviously connected with their astronomical observations and knowledge. On studying inscriptions in which calendar corrections are employed, evidence is gained, among other things, of three most important discoveries, striking examples of Maya astronomical learning: the length of the tropical year, the synodic revolution of Venus, and the lunation periods.

If according to modern astronomy the tropical year is 365.2422 days long, it is amazing to discover that the Maya sages had reached an extraordinary approximation, assigning it a period of 365.2420 days. If one compares the Maya achievement (Morley, *op. cit.*, 304–305; Thompson 1954: 158–161) with the computation implied by the Gregorian calendar (365.2425 days), it is clear that the pre-Hispanic astronomers had attained a more than exceptional precision (Figure 5). This notable discovery took place in and functioned during the Classic Maya horizon. Its origins are probably to be found in Copán around the sixth century

Figure 5. Glyphs from Stela A at Copán, after Teeple. Aside from the three dates included in the long inscription which appears on this stela, these glyphs, which recall those of the supplementary series, permitted Teeple to discover the formulae employed by the Maya to adjust their computations of the solar with the tropical year. See Teeple, *op. cit.*, 70–75

A.D.[8] Even if we knew nothing more about Maya astronomy than the computations of the year, this would force us to recognize the Mayas' obsessive interest in formulating a science of time.

Regarding their knowledge about the cycle of *Noh ek* (Venus), "the great star," the *Dresden Codex* reveals, on pages 46–50, the preservation of a sort of Venusian calendar of considerable precision. The sages and priests found it important to correlate what is today called the synodic revolutions of Venus with their computation of the year and with the *tzolkin,* or 260-day count. From the point of view both of ceremonial organization and of astrological prediction, it was necessary to discern the moment in which the three different computations (Venusian, solar and *tzolkin*), with their corresponding deities, were to coincide in a "resting place" to later reinitiate their perambulations with their burdens of unending time.

The synodic revolution of Venus has an average duration of 583.92 days. The original computation of the Maya—584 days—inevitably caused them to incur an error of forward displacement. A study carried out by John E. Teeple, concerning the pages of the *Dresden Codex* in which appear the cycles of Venus, reveals that, with a small margin, the Maya astronomers suppressed the discrepancy. The procedure which they followed can be synthesized in the following way: repeated observations showed that if, after sixty-one Venus years, they subtracted 4 days, the total of those

8. See Teeple 1930: 70–80. Concerning the degree of exactitude achieved by the Copán Maya sages in their calculations on the duration of the tropical year, Teeple (*op. cit.,* p. 74) offers the following comparative list:

Present duration of the year	365.2422 days
Duration of year about A.D. 600	365.2423
Julian year:	365.2500
Gregorian year:	365.2425
Maya year at Copán	365.2420

elapsed, during as many Venus years, could be made divisible by 260, that is, by the number of days in the *tzolkin.* In this manner a first form of correlation could be realized between both counts. But the 365-day year continued to offer problems.

The new corrections adopted must have resulted from equally multiple observations: it was necessary, every five cycles, to introduce a new modification of eight days at the end of the fifty-seventh revolution. Without forfeiting the correlation with the *tzolkin,* they succeeded in obtaining the interconnection they sought with the solar year. The interposed corrections certainly implied an error in the computation of the synodic revolution of Venus, but this consisted only of 0.08 of a day in every 481 years. This achievement very definitely does honor to the chronological endeavors of the Maya.

In the *Dresden Codex* (pages 51–58), there is an indication of some computations in relation to the cycle technically designated as a lunation period, *i.e.,* a complete revolution of the moon around the earth. For ceremonial and astrological reasons, from the Classic era the Maya were interested in knowing with precision the length of the lunar cycles. Proof of this exists in the stelae on which, when carving a date, they recorded the lunar age as well as the position of the corresponding month in their lunar half year.

Various investigators, among them Teeple, have tried to clarify the information available in the so-called "lunar table" of the *Dresden Codex* (Teeple 1930:86–98; Thompson 1960:230–246). Among their most important findings can be cited the following: the Maya related the lunar cycles with the 260-day count. In this way they established units of measurement that made it possible to calculate lunar variations at different moments in the distant past. Recording on repeated occasions the moments within series of double *tzolkins* (2 × 260 days) in which, as a result of the moon's position, there had occurred eclipses of the sun, they succeeded in elaborating a table, valid for predicting 69 possible eclipses in periods of approximately 33 years (11,960 days = 46 cycles of 260 days). Finally, by means of this table, they were also able to reduce the discrepancies among the new moons forecast by their computations and the new moons of astronomical actuality; this is proved by the fact that in a series of 405 successive lunations the error was never greater than one day. This, comments Morley, is "a truly colossal achievement for any chronological system, whether ancient or modern" (1947:289).

This brief description of facts about Maya chronological and astronomical knowledge already points to the enduring fascination of this people in the exploration of the mysteries of the universe, discerning the significance and measurement of its cycles. No other ancient culture was able to formulate, as they did, such a number of units of measurement and categories or so many mathematical relations for framing, with a tireless desire for exactitude, the cyclic reality of time. By mentioning some of the achievements in the fields of astronomy, chronology, and mathematics, my aim has been to point out the greatest achievements of Maya wisdom concerning the measurement of time. This will facilitate the understanding of the question here to be investigated: What significance had time and the computation of its cycles within Maya mythology and the world view of its priests and wise men?

CHAPTER II

MAYA SYMBOLS AND EXPRESSIONS OF TIME

The richness of Maya thought about time, besides the strictly calendrical and chronological knowledge, becomes evident through the terms, glyphs, concepts, and texts related to this theme. The sole enumeration of the principal symbols and concepts having temporal connotations is in itself eloquent:

a) Those which express periods or cycles of time: *kin* (day), *uinal* (month), *tun* (year), *katun* (twenty years), *baktun* (four hundred years), and so forth.

b) The numerical glyphs and their variants.

c) Glyphs of the twenty-day series.

d) Glyphs of the eighteen months and of the five days at the end of the year.

e) Glyphs of the cosmic directions within their temporal relationship.

f) Symbols and attributes of the gods who bear the burdens of time.

g) Glyphs of the divisions of the day and the night.

h) Glyphs of the patron deities and protectors of determined periods or cycles.

i) Expressions of strictly astronomical character as related to computations of cycles of the sun, Venus, the moon, eclipses, and so forth.

j) Symbolism of the fiestas, ceremonies, and rites, determined by the calendrical computations.

k) The late texts, especially those related to prophecies of the diverse *katuns*.

Though it may seem superfluous, it can be added that, for studying this accumulation of expressions so rich in temporal significance, there exist various categories of sources, three of which are of a totally indigenous origin:

a) The chronological inscriptions on the stelae and monuments starting at latest in the third century A.D. and present in more than ninety sites during the stage of florescence (A.D. 600 to 800) of the Classic period up to the moment in which stelae were no longer erected, toward the end of the tenth century. Sources are also other forms of symbolism of the Classic period and to a lesser degree of the Postclassic.

b) The three extant Maya codices of pre-Columbian origin (from the Postclassic period).

c) The various late writings redacted in Maya languages, but in the Latin alphabet, by native sages and priests who survived the Conquest: the books of *Chilam Balam,* the *Popol Vuh,* and other texts.

Other sources of information are the works of the Spanish chroniclers, principally those of Diego de Landa and Diego López de Cogolludo as regards Yucatán, of Antonio de Fuentes y Guzmán in respect to Guatemala—as well as the *Historia* by Father Francisco Ximénez—and, finally, various geographic accounts of the sixteenth century and other writings among which mention only will be made of *Informe contra Idolorum cultores* by Pedro Sánchez de Aguilar.[1]

Supported, above all, by sources of indigenous origin we will seek possible forms of meaning in the theme of time throughout the evolution of Maya culture within the con-

1. Precise references to each of these sources will be given throughout the text as well as cited in the bibliography.

text of its mythology and religious thought. Obviously, the adoption of an adequate method of historical comprehension is required. This implies a strict adherence to the original sources as well as an avoidance, as far as possible, of attributions and concepts foreign to Maya mentality in order to perceive that which is native to it.

As a point of departure it is not necessary to repeat here the analysis, already made by various investigators, of a considerable part of the Maya categories or kinds of terms, glyphs, symbols, and concepts with temporal connotation. In each case throughout this work, there are offered proofs supporting the interpretations or inferences formulated. The adopted point of departure concentrates on the following question: Do there exist in the lexicon of the numerous languages of the Maya family one or various terms which in some way connote, in spite of foreseeable differences, an idea generically akin to that meant by our word *time?*

Recent studies by various linguists, specialists in the comparative analysis of Maya languages, offer a first form of reply. Touching on the problem of origins and differentiation of the Maya peoples from the point of view of the analysis of their respective tongues, Norman McQuown has compiled a list of linguistic elements which are identical or closely related in the diverse tongues integrating the Maya family. In his opinion, those morphemes appearing with the same meaning in the totality of Maya languages "were present in the original community, before it began to disintegrate," as much by geographic dispersion as by its own evolution (McQuown 1964:77).

In the assembled list of materials from twenty-five languages of this family, a total concordance ("cognates" according to linguistic terminology) has been found regarding 218 vocabulary items. Among those connoting religious beliefs exist two particularly interesting cases: *kuh,* expressing the idea of "something sacred or divine," and *lab,* signifying "bad spirit" or "nefarious influence." Returning to the question previously posed about the idea of time, the reply is that, in all the Maya family, there exists a concordance of the related word or cognate *q'iing* or *kinh,* the *kin* of the Yucatec Maya, consistently signifying sun, day, and time (*ibid.*:78).[2]

The fact that this term is present—not only at the time of the Conquest, but in our own time in the vocabulary of such separated groups as the Yucatec Maya as compared to the Quiché, Cakchiquel, Mam, Pocomán, and others of the Highlands of Guatemala as well as the Tzotzil and Tzeltal of Chiapas, themselves groups considerably different from the Maya communities of Honduras—is proof of the ancient origin of the semantic complex "sun-day-time" connoted by *kinh.* Furthermore, there exist various well-known glyphs which will be examined and which symbolically represent the meaning of *kinh* (especially as sun-day). These glyphs appear in the late Maya codices—and also in the Classic stelae and monuments of the Guatemalan Petén, of Yucatán and Campeche, of the Usumacinta river basin, of various places in Chiapas and of sites such as Quiriguá and Copán.

2. McQuown, Norman A. *Ibid.*:78. In order to facilitate the pronunciation of this term of probable proto-Mayan origin, we will modify the spelling adopted by McQuown. Instead of *q'iing,* we will write *kinh,* using *k* to represent the glottalized postvelar sound, and *nh* for the nasal velar. Since in some Maya languages the *i* in this morpheme is probably short, it does not seem necessary to use two *i*'s to indicate a long vowel. On adopting the spelling *kinh,* its relationship with the *kin* of the Yucatec and with the *kih* of the Quiché becomes more obvious.

See also the work of Terrence S. Kaufman, "Materiales Lingüísticos para el Estudio de las Relaciones Internas y Externas de la Familia de Idiomas Mayanos," in *Desarrollo cultural de los mayas,* pp. 81–136 (particularly pp. 103 and 111). Also, Mauricio Swadesh, "Interrelaciones de las lenguas mayas," in *Anales del Instituto Nacional de Antropología e Historia,* 42, Mexico, 1961, pp. 231–267.

From the archaeological point of view they also corroborate the ancient origin of this primordial expression about time.

In the following pages, the main variant glyphs representing *kinh* will be treated at length. At this point it can be stated that the most common is the one simulating a flower with four petals (Figure 6). This, apart from being found in innumerable inscriptions of strictly calendrical content, also appears alone or with distinct affixes in many other nonchronological texts, inscribed since the beginning of the Classic era. As Thompson has demonstrated in his catalogue of Maya hieroglyphs, the *kinh* sign ranks among "the four most frequent glyphs of noncalendric use," aside from its frequent presence in chronological contexts (Thompson 1960:22). Apparently, this constitutes a first form of proof of the numerous relationships linking *kinh* (sun-day-time) with symbols probably of religious, mythological, or ceremonial content in the multiple inscriptions on which it appears.

But to penetrate the rich complex of meanings implied by *kinh,* before continuing the study of glyphs representing it, it will be necessary to make a brief semantic analysis of the term itself and of a few words derived from the same root in various Maya languages. The primary meaning of *kinh* is apparently "sun." From the time it rises in the east (*la-k-kin*) in Yucatec Maya: the "accompanying sun") until it sets in the afternoon (*chi-kin:* "the sun in the mouth," or "the devoured sun"), the sun's travel creates the day, marking its duration and existence. Therefore, "day" is simply a presence or cycle of the sun. Semantically, the relationship is obvious.

The sun does not rest, however. When it is apparently "devoured" in *chi-kin,* its setting, it goes into the lower world, crosses it, and triumphantly is reborn. Throughout the cosmic ages, which are the "suns" in the cosmogonic text of the *Popol Vuh, kinh* (as sun) in the end always re-

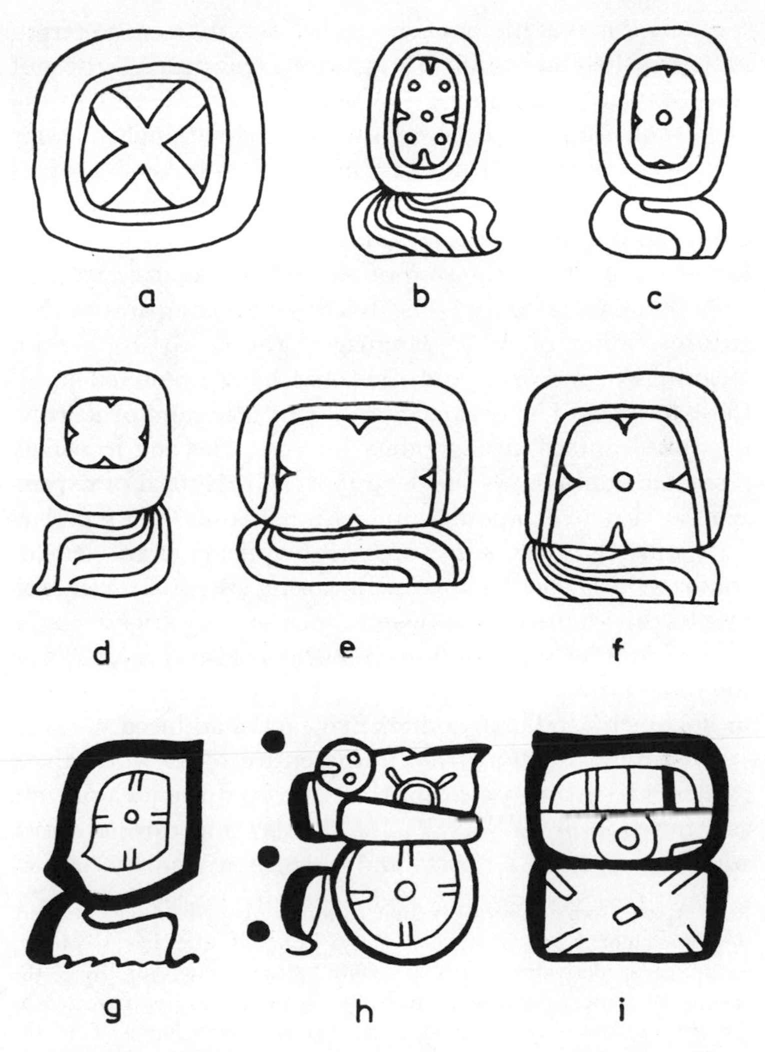

Figure 6. Kinh glyphs: the four-petaled flower in the carved inscriptions and the codices. *a*) Uaxactun 26; *b*) Copán I; *c*) Foliated Cross at Palenque; *d*) Copán M; *e*) Quiriguá P; *f*) Quiriguá I; *g*) *Dresden Codex* 61; *h*) "Brilliant Sun" with the *te* affix, *Dresden Codex* 72; *i*) *Kintun,* "Burning Sun," *Madrid Codex* 34 (*Source:* THOMPSON, *Maya Hieroglyphic Writing*)

surges and makes life possible. Its cycles only seem to terminate. On their stelae, the Maya priests computed "scores of suns" dating back hundreds of millions of years and, at the same time, forecast future cycles. If in their thought the day was a solar presence, time was the limitless succession of all solar cycles. Thus *kinh* spontaneously acquired its most ample meaning: duration that cannot be expressed because it has no limits, time, the sum of all possible solar cycles.

As demonstrated by the already-cited comparative linguistic studies of Maya languages, the word *kinh,* with meanings of sun, day, and time, has been preserved in all those tongues. The permanence of this term and of its complex of meanings throughout many centuries and in widely separated zones shows the deep roots of this form of expression so rich in temporal connotations. Some of its derivatives and other words in which *kinh* appears as an element confirm the ancient temporal meaning essential to it. Examples can be cited of expressions such as *ubay kin* ("when"), *ukin* ("time of something") in Yucatec Maya, *k'hir* ("pass the time") in Quiché, *k'ij yu'nac* ("today, the present time") in Pokomchí, and many others that can be adduced.[3]

We know, therefore, that in the entire Maya world there existed a term and a concept that conveyed the idea of time as a function of the solar cycles, the day and the sun itself, whose never-ending risings and settings govern the fate of

3. Apart from the sun-day-time connotations already mentioned, the *kinh* morpheme in various Maya tongues also has secondary meanings or derivatives such as "fiesta" (festal day), "destiny" (the fate or attribute, fortunate or unfortunate, of the day as indicated by the *tzolkin,* the astrological calendar). It is interesting to note the parallelism which exists, to a certain degree, between the semantic evolution of *kinh* and of the Nahuatl word *tonalli.* The morpheme *tona* signifies "give light, give heat." From it is derived *tona-tiuh,* "he who goes along giving light" (the sun). *Tona-lli,* as a result, is the day. *Tonalli* also connotes the destinies inherent in temporal cycles, in terms of the *tonal-pohualli,* or 260-day count, equivalent in the Mexican Highlands to the *tzolkin* of the Maya.

all that exists. So much can be gained through linguistic analysis. Now, let us consider that which can be deduced from the glyphs and symbolism, beginning with the Classic period.

The stelae bearing calendrical inscriptions, in accordance with the Long Count system, are our primary sources. In these, as stated in the preceding chapter, time is computed, according to the position of the glyphs, to indicate: the *baktuns* (360 × 20 × 20 days), the *katuns* (360 × 20) the *tuns* (360), the *uinals* (20 days), and finally the *kins* or days. It is natural, therefore, to find the *kinh* glyph in one or another of its variants in these inscriptions. But what do we know, through the research of the specialists, about the symbolism in the varying forms of this glyph? Eric Thompson writes:

> The symbolic form of this glyph is the main element of the glyph of the sun god. It also appears frequently as an identifying attribute on the forehead, the earplug, or the headdress of that deity, and it is also the principal element of the month sign *yaxkin*.
>
> The glyph resembles, and in all probability represents, a four petaled flower. It seems very probable that this is a conventionalized picture of some species of plumeria. The plumeria is a symbol of procreation. . . .
>
> This *kin* element has a postfix, a streamerlike arrangement, which, it has been suggested, is the beard of the sun god and which is sometimes called the tail. . . .
>
> The most easily recognized and perhaps the commonest variant of the head form is that of the sun god himself. The characteristic features of the sun god are: a squarish eye with squarish pupil in the top inner corner and with a loop, often with two or three circlets attached, which encloses the eye on the sides and bottom; a prominent Roman nose; the central incisors of the upper jaw filed to the shape of a squat

> tau; often a fang projecting from the corner of the mouth; and a hollow on the top of the head. In the glyphs the *kin* tail is usually present beside or below the head.
> The second variant of the head is that of an animal. The nose becomes a snout, often with a small scroll on it. The eye is still squarish, but the pupil moves to the center of the eye and becomes a short crescent, the ends of which are sometimes joined across the top by a straight line. A shell pendant hangs from the earplug, and on the cheek there is frequently an irregular crescent, the horns of which point toward the ears. . . .
>
> 1960:142

The traits present in these forms of symbolism, characteristic of *kinh,* indeed manifest the richness of their connotations and at the same time the precision of the analysis given to them on the part of Maya scholars. To complete the listing of the variants of this glyph, we will only state, following Thompson, that the zoomorphic head to which he referred seems to be a stylization of that of a jaguar or of a dog. In more than one case there are also variants with the figure of a simian which, by its attributes and other forms of connotation, is related to the solar symbolism. "These variant forms (it is worth reiterating) well illustrate how deeply the hieroglyphs are rooted in mythology" (*ibid.:* 143).

Already noted is the abundance of calendrical and noncalendrical inscriptions in which variants of the *kinh* glyph appear. They date from the first stelae of the Classic period and later from other symbol complexes as well as from the codices, two of which (the Paris and the Madrid) probably were painted not long before the Conquest. These sources permit us to affirm that, as with the concept and term of *kinh,* its diverse forms of glyphic representation were also an ancient patrimony shared by most groups of the Maya

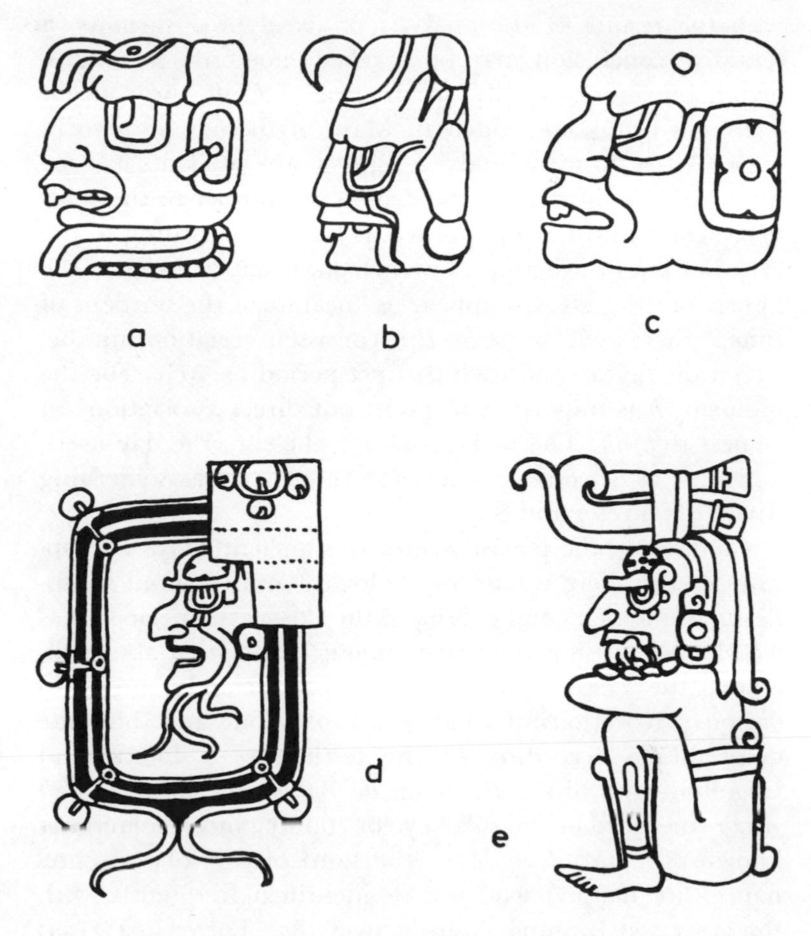

Figure 7. The Solar God, a hieroglyphic variant of *Kinh. a*) Yaxchilan L; *b*) Piedras Negras 14; *c*) Quiriguá, Structure I; *d*) *Dresden Codex* 55a; *e*) *Madrid Codex* 71a

family. As an illustration are presented (Figure 7) various examples from the inscriptions of Yaxchilán, Piedras Negras, and Quiriguá as well as from the Dresden and Madrid codices.

If the semantics of *kinh,* sun-day-time, can now be related with the results of the analysis of its glyphic variants, a tentative conclusion may be reached: not only the hieroglyphs but more broadly, in the case of *kinh,* the concept itself was firmly embedded in Maya mythology and world view. Apart from secondary aspects, the variants of the *kinh* hieroglyph point at the symbolism proper to the solar deity, among others that known as *God G,* according to the classification of Schellhas. Further on, when treating of the figures of the gods that appear as "bearers of the burdens of time," there will be seen the consistent relationship between the deities and each distinct period or cycle. For the moment, it is important to point out direct associations in respect to *kinh.* The maker of days, the sun, the day itself, and time as a whole, are already thought of as something divine (Figures 7 and 8).

Confirming the persistence of this ancient Maya conception of *kinh* there is later mythological and religious speculation of the sages and priests of the Postclassic period. It is well known, as shown by Seler among others, that above all, among the Yucatec Maya, *kinh* appears closely linked to various advocations of what seems to be one and the same deity.[4] Thus according to the testimony of Landa and Cogolludo and of the *Relación de Valladolid,* he is *Kinich Ahau* (the Lord of the solar eye or countenance), sometimes known as *Kinich Kak Moo* (the Lord of the solar countenance, fire macaw) who is also identified frequently with the great god Itzamná. (Landa 1938:187; Tozzer 1941:153; Cogolludo 1954:I:352). It may be added that there exists

4. Concerning Itzamná, Eduard Seler discusses the multiple inclusions of *kin* among the attributes of the various titles of this same divinity in "Ueber die Namen der in der Dresdenen Handschrift angebildeten Maya Götter" (1902–1923:I:378–379). See also Rudolph Anders, *Das Pantheon der Maya,* pp. 320–321 (concerning the relations of *Kinich Ahau* and *Kinich Kak Moo*); pp. 303–309 (concerning Itzamná).

Figure 8. Kinh, sun-time, deity, in the symbolism of the Classic period and in the codices. *a*) The sun in its relationship with the earth with glyphs of *caban.* Temple of the Sun, Palenque; *b*) The face of the sun in the table of the eclipses, *Dresden Codex* 56a; *c*) Deity with a sun glyph on its forehead. It emerges from the jaws of a crocodile, Copán; *d*) The sun glyph under a strip of celestial symbols. It is about to be devoured by the Earth Monster, *Madrid Codex* 25a

more than one source according to which *Kinich Ahau Itzamná* was worshiped as the son or solar manifestation of the supreme and only god, *Hunab Ku.*[5]

These relationships, valid at least for the Postclassic Maya of Yucatán, could be amplified taking into account the evidence of the codices, especially that of the oldest, the *Dresden*. On several of its pages are shown divinities with solar attributes and even with the glyph of the sun. As noted by Seler, they constitute a kind of rich theological complex with antecedents in the inscriptions and symbolism of the Classic period. Writes Seler:

> The second page of the *Dresden Codex* exhibits a god who on the forehead and hieroglyph carries the sign that Rosny has already recognized as that of the sun and as an emblem of *kin*. For this reason, Schellhas in his study of the hieroglyphic signs of the deities in the *Dresden Codex* identifies him with the *Kinich Ahau* of the historical sources. The solar god is also the god of war. Thus we see in the upper part of page 26 of the Dresden a priest as a yearbearer [actually *Chac,* god of rain], with an animal head instead of that of the god and a symbol displaying traits of a *balam* or jaguar (Figure 9a). . . .
>
> 1902–1923:I:378

Further on Seler comments on some pages of the *Madrid Codex:*

> On page 20, on referring to the year IX, were Itzamná should appear, on the lower left side, we see the god that bears the *Kan* glyph. He represents Itzamná but with closed

5. See the *Relación de Valladolid* which refers to the only god having the name of *Hunab ku* and *Zamana* (Itzamná), in *Colección de documentos inéditos relativos al descubrimiento, conquista y organización de las antiguas posesiones españolas de ultramar,* second series, vols. 11 and 13, Madrid, 1898–1900, Book II, p. 161.

a

b

Figure 9. Representations of *kinh* in his relationship with other deities: *a*) One of the *chacs* with a sun-jaguar on his back, *Dresden Codex* 26a; *b*) On the left is seated a solar deity under the sign of *kan*. Facing it appears the sun-god, Itzamná (after Seler) *Madrid Codex*, 20c

eyes. In front of him, on the lower right side, we find the solar god with a sort of mask in the form of a bird showing the eye of the old god Itzamná. This is an extraordinary concordance with what Landa tells: in the years IX, apart from venerating the patrons corresponding to the year, they also feted the Lord of the solar countenance, the god *Kinich Ahau Itzamná* (Figure 9b).

Ibid.:386

What Seler affirms regarding the gods, within the essentially temporal context of the calendrical computations, is only a sample of the great nucleus of mythological interrelationships in which are present the symbolism and the ancient concept of *kinh*. These examples and others that could be cited[6] prove the survival in the Postclassic period of the age-old idea of time as something coming from the divinity and somehow part of its very being.

One last kind of evidence is offered by certain texts in Maya, writings made in post-Conquest days. These accounts are dependable notwithstanding obvious cultural elements present in them derived from the Central plateau and, later, from the Christian teachings of the missionaries. Particularly important are various passages of the books of *Chilam Balam,* especially the texts of chronological and prophetic content such as those dealing with the series or "wheels of the *katuns*." There, on describing the fortunate or unfortunate character of the different periods, there often appears a reflection on the divine nature of *kinh,* day-sun-time. Thus, for example, in the first prophetic wheel of a series of *katuns,* which appears consistently in several of the books of the *Chilam Balam,* the solar deity bears the title "Countenance of the Sun, Fire Macaw" (*Kinich Kak*

6. See, among the more obvious examples, *Dresden Codex,* 47c, 55 and *Madrid Codex,* 37, 75–76.

Plate I. A cylindrical, polychrome tube of clay showing the face of *kinh* who emerges from the fangs of the Earth Monster. There also appear shells, fishes and a bird. Its provenience is the Temple of the Cross, Palenque. (*National Museum of Anthropology, Mexico City*)

Moo), presiding over and governing the period of time of a 6-*Ahau katun*. One of these texts clearly showing its relationship with the ancient hieroglyphic computations, reads as follows:

> This is the word and the order
> displayed by the signs
> of the house of 6-*Ahau katun*.
> This is the word and the order
> in which come the time
> and the years of the *katun*.
> In *Uucil Yabnal*
> (in the place of his "seven waters")
> stands the seat of the 6-*Ahau katun*.
> *Kinich kak moo,*
> Countenance of the Sun, Fire Macaw,
> at Uxmal;
> his face will be in the heavens
> during this reign
> that will be of shameless stares
> and of foolish perception.
> Sadness will come
> on the arrival
> of the usurpers of the mat,
> of the usurpers of the royal seat. . . .
>
> BARRERA VÁSQUEZ 1948:117

The face of the sun, *kin ich,* in the role of *kak moo,* Fire Macaw, bears the inherent reality and destiny of 6-*Ahau katun* in its burden of time. Fateful is this series of omens because almost all of them refer to the imminent destruction of the ancient way of life. We will find in them again and again the various temporal symbols in unending association with the divinity. In almost all cases it is the very priests of the *kinh* cult (the *Ah Kin*) who speak, and reveal the burden of time which each period carries: "Thus spoke

Plate II. The head of *kinh* in carved limestone. It seems to have been part of the staircase of the hieroglyphs in Copán. (*University of Philadelphia Museum*)

the great *Ah Kin,* priest of the cult of the sun, *Chilam Balam,* interpreter, when he inscribed the signs on the face of the 8-*Ahau katun* . . ." (*ibid.*:117).

The allusions to *kinh,* time and deity, are constant:

> Under the might of *Ah Uuc Kin* (Lord 7-Sun) . . . (*ibid.*:99). It is the seat of the 12-*Ahau katun, Yaxan Chuen,* Great-monkey-craftsman [a solar title]. It is the countenance he will display during his reign in the heavens. There will be great sages, great sorcerers. That which is in the heavens will come forth on 12-*Ahau.* It will be the *Ah Kin,* priest of the solar cult, who will seat upon the mat and the throne wearing the jaguar mask (*ibid.*:112). [A new allusion to the sun, as we have seen on dealing with the codices.]

Finally, Itzamná, in his relationship with the sun, is not absent in the prophecies of the *katuns:*

> This is his word,
> *Kinchil Cobá*
> "Bird of solar face and eye,"
> this is the seat of 13-*Ahau katun*. . . .
> It utters its word,
> it shows its face to say its word
> this *katun*
> that has the face of Itzamná. . . .
>
> *Ibid.*:121–122

The metaphors and mythological allusions found in these texts concerning *kinh,* as sun and time, maintain a surprising similarity, if not an identity, with the symbolism present in the codices and in the inscriptions of the Classic period. Thompson, in discussing the variants of the *kinh* glyph, has pointed out the following as the most frequent and important: the old god with the characteristic solar eye, the stylized flower with four petals, the jaguar mask, and the monkey face. These variants, appearing in the Classic inscriptions and symbolism, are also found in the *Dresden Codex* and in the later manuscripts known as the Paris and Madrid codices. Finally, as already shown, the god of the solar eye, the face of the great craftsman monkey and the jaguar mask maintain the same symbolic character related directly to *kinh* through the *katun* prophecies in the books of *Chilam Balam*.

Admittedly, both the codices and also the late Yucatec texts contain other elements absent in Classic symbolism. To give an example, Itzamná, as title of the solar deity, appears probably as a consequence of subsequent forms of religious syncretism. It is remarkable, however, in spite of differences, to find a persistence of similarities and identities, all centering around time and related to the various meanings of *kinh,* present through the ages, even today, in the Maya world.

Two conclusions may be reached, at this point, as a result of the confrontation of the different sources we have studied: the first refers to the particular connotation which, since Classic days, the complex of symbols related to *kinh* exerted over the thought of the Maya. *Kinh*—sun-day-time—was not an abstract entity but a reality enmeshed in the world of myths, a divine being, origin of the cycles which govern all existing things.

Many are the faces of *kinh,* but its essence is always divine. Time permeates all and is limitless. Thus, the priests computed millions of years into the past and as many others into the future. Time is—and we will deal with this in the following chapter—an attribute of the gods: they carry it on their backs. In a word, *kinh* appears, as the heart of all change, filled with lucky and unlucky destinies within the cyclic reality of the universe and most probably inherent to the essence of divinity itself.

The second conclusion has already been noted: from the beginning of the Classic period to post-Conquest times, when the "wheels of the *katuns*" were transcribed and continued, for longer than a millennium and a half, not a little of Maya time symbolism has perdured. This has occurred in spite of obvious innovations and outer influences through the centuries. Maya computations of and concern for time, doubtlessly reveal a penetrating mind. Its study, notwithstanding the difficulties involved, can open a door for us modern men of a unique way of reasoning.

In a post-Conquest collection of chants from the town of Dzitbalché in Yucatán, studied by Alfredo Barrera Vásquez, is found a hymn that contains a solemn and profound declaration of what the primordial reality of time may have meant to priests and sages. Though the date of the composition of this hymn is undoubtedly late, the ideas expressed reflect ancient tradition.

> Only Thee
> do I trust entirely,
> here where one dwells.
> For thou, oh great kin,
> providest that which is good,
> here where one dwells,
> to all living beings.
> Since Thou abidest to give reality to the earth,
> where all men live.
> And Thou art the true helper
> who grants that which is good.
>
> BARRERA VÁSQUEZ 1965:46–47

CHAPTER III

TIME AS AN ATTRIBUTE OF THE GODS

We have seen that the Maya attributed a divine nature to *kinh,* sun-day-time. The day, and all the cycles, owed their being to the old face with the solar eye, the ascending fire macaw, the jaguar deity or the dog, the two latter symbols of the occultation and voyage of the sun through the somber regions of the underworld. In his untiring coming and going through the paths of the universe, *kinh* brings with him attributes and influences belonging to the different periods and moments registered in the inscriptions and the codices. Throughout the great "suns" or ages of the world, all the days, the twenty-day periods, the years, the twenty-year periods, and the counts of all possible cycles—these all arrive with their varied messages, the nature of which man must foreknow in order to deal with their good or bad influences.

Maya sages, masters in the measurement of time, expressed their computations combining numbers with the glyphs of the days and months (*uinals*) within the Long Count. Later, in the Postclassic period, the same was achieved by means of the solar sign of *Ahau,* "the great lord," and its corresponding cipher through the cycles known as "wheels of the *katuns.*" Consequently, the twenty glyphs of the days and the eighteen glyphs of the *uinals,* as well as the numerals, are a primary key for penetrating the world of meanings time had for these ancient Americans.

As in the case of *kinh* itself, all numbers and glyphs were not abstract entities but the faces and supernatural personifications of the good and bad forces unceasingly interacting in the world. Thompson expresses it briefly:

> The days are alive; they are personified powers, to whom the Maya address their devotions, and their influences pervade every activity and every walk of life; they are, in truth, very gods.
>
> 1960:69

It would be impossible to study the pantheon or the theological thought of the Maya without devoting particular attention to the deities of each period and to their complex interrelationships in the chronological context. One wonders if it is not in this that the heart of the Maya conception of the divine and the cosmic must be sought. In the various attempts to identify and correlate the gods, therefore, a decisive step has been that of clarifying the meanings of these calendric glyphs.

Regarding the numbers, those from one to twenty, including the zero or symbol of completion, are fundamental. The vigesimal system of counting used by this people justifies the latter statement. Thanks to the aforementioned comparative linguistic studies we know that the names of the numbers in the various Mayan tongues stem from a common origin and in some cases even are identical (McQuown, *op. cit.*, 79; Kaufman, *op. cit.*, 113–114). This situation confirms the antiquity not only of the vigesimal system among the Maya, but also of their conceptualization of the numerical entities expressed by cognate terms and by a widespread acceptance of consistent glyphic patterns.

The analysis of glyphic variations in the case of the numerals is revealing. On one hand, and most frequently used, we have dots expressing units and bars representing "fives." Here exist elements worthy of taking into account: colors are meant to indicate different uses of the numerals. As companions or "bearers" of the days, they are painted red and those of the months or twenty-day periods, black. On the other hand, as noted by Thompson:

> The Maya with their mystical attitude toward numbers were not satisfied to use only bars and dots to depict them. In many texts, although rarely in *Dresden* and never in the other codices, numbers are expressed by portraits of deities, whose features or attributes are the key to the number thus portrayed.
>
> *Ibid.*:131

J. T. Goodman was able to identify the images of each one of these "gods of numbers," bearers of time (1889–1902: appendix). Their identification and the study of their relationships with the deities of the days, months, and other periods have brought to light important aspects of the intricate chronological systems and also something of what time meant for the Maya.[1]

It is the inscriptions, carved in stone, dating back to the Classic period, which best reflect the ancient philosophy about the interminable succession of godly faces and figures carrying the deified time cycles. Throughout the cosmic ages life was reborn thanks to *kinh*. Man recognized and thus approached the gods as bearers of the different periods: their faces were living portraits of time. The Maya sages thus dwelled in a universe tinged with mythological meanings and relationships. Every moment was the manifestation of forces, favorable or adverse, but always divine. As on a revolving stage, the gods of day and night, those of the

1. It does not seem to be a mere coincidence that the frequently cited Eric Thompson deserves not only to be considered a scholar in the field of Mayan hieroglyphic writing but also the first investigator of what he called "Maya philosophy of time."

months and numbers, the deities of all the cycles of time, were the actors in this universe. Their appearances and exits determined men's fates as they brought life and death with them. In computing periods of time, the priests endeavored to predict what would be the acts of the gods, to predict the nature of each deified moment.

To approach the core of this peculiar world of ideas it seems indispensable to recall first the most significant of the attributes of the main period-of-time gods, the *dramatis personae* in the universe of the ancient Maya. First of all, we will consider the *kinh* day-series, the twenty day-deities, afterward to be related to the gods of the numbers, months, years, and the scores of years, the *katuns*. In the case of the twenty day-deities, the calendrical designations of the Yucatec Maya will be discussed. These names are the most commonly used by students of the subject. Attention will also be given to the symbolism expressed in the glyphs and inscriptions. Though concordance does not always exist among all the terms expressing day-names in the Maya languages, one can affirm that the meanings of the various words designating a day or month are akin.[2]

1. *Imix,* the first of the days, connotes the earth-monster deity, the root from which all things spring. Among its symbols are the water lily, the head of a species of dragon or fantastic crocodile lacking the lower jaw, with a hanging protuberance for a nose.

2. *Ik,* wind and life, is a term and concept found among all Mayan groups. *Ik* introduces the god of rain.

2. See the table with the name of the days in various Mayan languages as presented by Thompson in *Maya Hieroglyphic Writing,* p. 68. In addition to affinities in connotations of several terms corresponding to the same day, there are seven days designated with cognates, that is, closely related terms, in the following languages: Yucatec Maya, Tzeltal and Tzotzil, Chuc (San Mateo), Jalaltec, Ixil, Quiché, and Pokomchí. Regarding the names of the months, see the corresponding table in *op. cit.,* p. 106.

3. *Akbal* is the darkness. It points at the underworld and is associated with the jaguar, since the night sun travels through those nether regions.

4. *Kan,* meaning ripeness, is the sign of the young maize god, the lord who brings abundance. The *kan* glyph is often found together with symbols of food offerings.

5. *Chicchan* is the celestial serpent as well as the four ophidians living in the heavens, at the four corners of the world, causing rainfall. Its various glyphic forms are reminiscent of the snake.

6. *Cimi,* as shown in its plastic representations, is the name-day of the god of death. The owl, omnipresent omen of death in Indian Central America even today, is closely related to *Cimi.* The personified form of its glyph is a skull.

7. *Manik* is represented by a hand which is at times associated in the codices with the divine Itzamná of the Yucatec. *Manik* in the Classic period was the day-sign of the god of the hunt.

8. *Lamat* is the sign of the lord of the "great star," that is, the planet Venus. It is represented by a cross-sected cartouche, each section displaying a small circle. In personified variants the celestial dragon bearing the four Venus circles is found in its place.

9. *Muluc* is symbolized by jade and water and is an aspect of the rain deities. When personified it appears as the head of a fish, perhaps the *xoc,* a large mythical fish.

10. *Oc* is presented as a dog's head. It guides, once more, the sun in its journey through the somber regions of the underworld.

11. *Chuen* makes its entrance in the day series as another aspect of the solar divinity. Disguised under the face of a monkey it appears as "the great craftsman," patron of knowledge and the arts.

12. *Eb* displays a face with a prominent jaw. In combination with the nineteenth day sign, *Cauac,* it evokes the

god who sends rains, drizzles, and mists harmful to the crops.

13. *Ben* is the day of the lord who fosters the growth of the maize stalk. He may also symbolize the growth and development of man.

14. *Ix* is a repeated appearance of the jaguar god in his relationship with the earth and the lower world. Its glyph though highly stylized includes spots of the skin of the jaguar.

15. *Men* introduces the aged face of the moon goddess. Thus she is portrayed in the ancient glyphs which indicate this day-sign. It is highly probable that she is to be identified with the ever-present Central American mother goddess, known under numerous advocations.

16. *Cib* is a day represented by a glyph in the form of a conch shell or a face again reminiscent of the jaguar god. It is probably related to the four *Bacabs* who supported the skies, also functioning as patrons of agriculture and beekeeping.

17. *Caban* is the youthful goddess of the earth, of maize, and of the moon: a young and at the same time aged deity. *Caban* was a lucky day for matchmaking, also auguring success for medical and commercial enterprises.

18. *Etznab* exhibits a sign, apparently that of the god of sacrifice: the sharp obsidian blade.

19. *Cauac* is the day of the celestial "dragons," deities of rain, lightning, thunderbolts, and tempests. The pictorial elements of its glyph are often enclosed in the head or body of the celestial "dragon." The representation of the latter often bears some similarity with the ophidian of the *Chicchan* glyph.

20. *Ahau* is the twentieth and last day-sign. This, the embodiment of the radiant presence of the sun, confirms that *kinh* is not only a divine countenance but that he him-

Figure 10. The glyphs of the days in the carved inscriptions and in the codices. These are examples of symbolic forms, faces, or personifications.

a) *Cimi:* Copán, breast-plate of a statue; Tikal, altar 5; *Dresden Codex* 12a; Landa.

b) *Oc:* Uaxactun, fresco G 1; Yaxchilan L; *Dresden Codex* 45a; *Madrid Codex* 45a.

c) *Eb:* Leiden Plate; Quiriguá C; *Dresden Codex* 12a; *Madrid Codex* 13b.

d) *Ahau:* Copán M.; Chichén 5; *Dresden Codex* 24; *Chilam Balam of Chumayel.*

(*Source*: THOMPSON, *Maya Hieroglyphic Writing*)

self is the Lord who encompasses the cycles of time[3] (Figure 10).

In a word, the days bear the attributes and countenances of the main deities of the ancient Maya pantheon. The sun, *kinh,* the lord, supreme ruler of time, appears six times, on the third, tenth, eleventh, fourteenth, sixteenth, and twentieth days of the series, with the masks of a jaguar, a dog, and a monkey, or as an eagle and lord under the sign of *Ahau*. The deity or deities of rain, as related to the wind or under the sign of ophidians or celestial dragons, are presented five times on the days that occupy the second, fifth, ninth, twelfth, and nineteenth places. The effigy of the young maize god, an ideal of beauty among the Maya, is shown twice, on the fourth and thirteenth days. The same number of times, on the fifteenth and seventeenth days, appears the young and old goddess of the moon who also makes corn grow and is venerated as the Lady of the Earth. The earth itself, the monster from which all is born, is precisely the first divine face in the day-series. Finally, we encounter the divinities of death, of the hunt, of the "great star" and of sacrifice on the sixth, seventh, eighth, and eighteenth days respectively.

If the faces of the day-gods are now compared with the deities of the numbers, we find, aside from the differences in their representations, several of the already-known figures, actors in the drama of time. Underlining the principal traits inherent to the numerals, one may offer the following condensed description:

Hun (one), *ca* (two), and *ox* (three) appear as three young faces. That of "one" belongs to the moon goddess, also

3. This brief description and the relationships noted with reference to the day-gods is based largely on the works of Eduard Seler, "Die Tageszeichen der Aztekischen und der Maya-Handschriften und ihre Gottheiten," *Gesammelte Abhandlungen,* vol. I, pp. 417–503, and of Eric S. Thompson's *Maya Hieroglyphic Writing,* pp. 69–93.

patroness of the day *Caban* and of the month *Kayab*. The "two" is the lord of sacrifice, related to the deity of the day *Etznab*. "Three," with the *Ik* symbol, is the deity of wind and rain, the same that appears on the day *IK*.

The faces of *can* (four) and of *ho* (five) are presented as old men. "Four" is *kinh,* the old sun, related to the day *Ahau* (Figure 11). "Five" is the numen of the interior of the earth, known also as *Mam,* god of the *Imix* day.

Uac (six) and *uuc* (seven) both display a flat nose. Six, related to God B, is the lord of rain and storms (Figure 11). Seven is the jaguar god, divinity of the lower world, with the symbol of night.

Once more young faces are encountered in *uaxac* (eight) and *bolon* (nine). Eight is the god of maize. Nine is *Chicchan,* the serpent god (Figure 11).

Lahun (ten) has the countenance of the god of death. *Buluc* (eleven) presents, as its characteristic sign, that of *Caban* the earth. It is the same god that governs the day *Manik,* of deer and hunting. He is the lord of forests. *Laca* (twelve) is another youthful-faced god with the heaven sign, closely associated with the planet Venus. *Oxlahun* (thirteen), apart from appearing sometimes as the sum of the attributes of 10 and 3, is shown as a water deity resembling the one presiding over the day *Muluc*.

The next numerals, from 14 to 19, do not conceal a fusion of traits, variants of the face of 10 with elements corresponding to those of the deities from 4 to 9. Finally, the sign of completion, "the zero," *lub,* means the end of the journey, place where the burden rests. It bears symbolic forms such as a shell or a hand as an attached element. Its personification is a face with traits peculiar to the god of the death.

The series of the eighteen months is, in an equal manner, a parade of faces, many of them now familiar to the reader, although as month signs they will appear with vari-

Figure 11. Deities of the numbers.

a) The number 4: at Copán 15, Foliated Cross, Palenque, Halakal 1.

b) The number 6: at Piedras Negras 12, Quiriguá A, Palenque 36 G 1.

c) The number 9: Yaxchilan L 48, Piedras Negras L 3, *Dresden Codex* 70.

(*Source:* THOMPSON, *Maya Hieroglyphic Writing*)

ants. Instead of describing the month signs in strict numerical order, from one to eighteen, they will be grouped in the following résumé under the attributes of their patron deities. Thus we will examine successively groups of months connected with the sun, water, the moon, "the great star" and other celestial bodies, the hunt and the earth.

The solar god, with his symbols of the jaguar and the mat indicative of his sovereignty, makes his entrance in *Pop* (straw mat), the first of the months. The same jaguar, traveling the underworld with the symbol of the black region, follows immediately in *Uo* (a small black frog), the name of the second month. In the sixth, *Xul,* the sun assumes the appearance of a dog with the characteristic tail sometimes included among the *kinh* elements (Figure 12). The term *Xul,* meaning "end," suggests the completion of the day when the sun enters the nether region. In *Yaxkin* (first or new sun, also time of drought), the seventh month, the solar god wears his most typical visage—that of an old man. Much later, in *Kankin* (yellow sun), the fourteenth month, the sun again takes on the face of a dog. Finally, in the sixteenth month, *Pax* (drum), the jaguar or flat-nosed god suggests the raising sun in close relation with rain.

The deities of the water that comes from the heavens are also faces and attributes for seven more months. At times with the sign of jade or of water, at others with that of the *moan* bird or in relationship with corn, fish, or the celestial ophidians, the god and the gods of rain exercise their influence on the fourth, fifth, eighth, thirteenth, fifteenth, sixteenth, and eighteenth months. *Zotz* (bat), the fourth, has as its patron the mythical fish *Xoc* which is typical of the personified glyph of the *Muluc* day intimately connected with the rain deities (Figure 12). *Zec* or *Tzec* (etymology uncertain), the fifth, is presided by a young god and its glyph exhibits the sign of *caan,* "the sky." During this month ceremonies were held honoring the Bacab patron of beekeep-

Figure 12. Glyphs of the months in the carved inscriptions and codices. The persistence of forms is remarkable, passing from the Classic, through the Postclassic, to Landa's sixteenth-century compilation.

a) *Zotz':* Copán 6; Foliated Cross of Palenque; *Dresden Codex* 47a; Landa.

b) *Xul:* Tikal, altar 5; Palenque 96; *Dresden Codex* 63b; Landa.

c) *Moan:* Yaxchilan L.; Quiriguá G; *Dresden Codex* 48b; Landa.

d) *Kayab:* Quiriguá K., Flores 2; *Dresden Codex* 61b; Landa.

(*Source:* THOMPSON, *Maya Hieroglyphic Writing*)

ing. The eighth, *Mol* (to gather), displays the jade or water sign. *Mac* (to close), the thirteenth month, brings again the symbol of the *xoc* fish, and is also related to the god of the *IK* day, numen of wind and rain. *Moan* (the Moan bird), the fifteenth, indicates the presence of water (Figure 12). *Cumkú* (god of the corn bin), the eighteenth month, has as its symbol a celestial "dragon" monster, portraying thus its connection with the all-important divinities of the waters.

The goddess of the moon makes two appearances, on *Ch'en* and on *Kayab,* the ninth and seventeenth months. *Ch'en* (a well), has in the personified variant of its glyph the lunar figure or the goddess coming out of the moon. *Kayab* (unknown etymology), is represented by a turtle's head. (Figure 12). Its patroness is the young lunar deity related to childbirth and medical practices.

A month under the custody of Venus, "the great star," is *Yax* (new or green), the tenth of the series. The head of the so-called Venus symbol often appears as its glyphic variant.

The mark of *Zac* (white), the eleventh month, seems to be a reptilian or a batrachian head. Its advocate is an undefined god linked with the heavenly bodies.

Finally, there are two more months on which the divinities of the hunt were revered. One is *Zip* (the name of the Yucatec god of hunting), the third in the monthly series. The other, *Ceh* (deer), occupies the twelfth position. A hunters' feast was held during this month in honor of *Ah Ceh,* the god of the deer. A certain association with Venus is indicated by this month's glyph.

Besides the 18 "months" of 20 days ($18 \times 20 = 360$), one has to consider the five days remaining at the end of the 365-day year, the *Uayeb* (unlucky), as they were called by the Yucatec Maya. The glyph of the *Uayeb* 5-day period is a year sign with a special prefix. These unlucky days fall under the influence of the Lord of the Earth.

To these deities—seen in the inscriptions of the Classic

Figure 13. Variants of the glyphs of *katuns, tuns,* and *uinals* on the carved inscriptions and in the *Dresden Codex.*

a) *katuns:* Copán J. *Dresden Codex* 61, Quiriguá F. Piedras Negras L 3.

b) *tuns:* Quiriguá J. *Dresden Codex* 61, Pusilha O. Naranjo HS.

c) *uinals:* Copán HS, *Dresden Codex* 61, Yaxchilan L. Sacchana I.

(*Source:* THOMPSON, *Maya Hieroglyphic Writing*)

period as personification of the days, the numerals, and the months—should be added many other symbols, also of divine character in relation with specific chronological computations. For our purpose it is sufficient to affirm that divine traits and influences were also attributed to other periods of time, generically conceived such as *kinh* (day), *uinal* (month), *tun* (year), *katun* (twenty-year period), and *baktun* (360 × 20 × 20 days)[4] (Figure 13).

The preceding description of all these temporal elements carries us one step forward in our study. On dealing with *kinh* in general it became clear that the Maya conceived of time in close association with the solar deity, something divine in itself, limitless and ubiquitous. Proof of this has been given by stelae inscriptions bearing computations placing precise moments of *kinh* at points as far back as millions of years. Glyphs in the ancient codices as also the Colonial Maya texts confirm this continuous obsession for knowledge and prevision of the ever-changing reality of *kinh,* the divine sun-day-time. Now, we know something more: all the moments of time—the days, months, and years—are arrivals and presences of divine faces. They all successively come and go, letting their influences be felt, unceasingly determining life and death in the universe. Each moment is not only the presence of one god but the sum total of many presences. The deities of the numbers, those of the days, months, years, and other time measurements, come together in different points of arrival throughout the cycles. The resultant of their forces colors reality with multiple tints. Such was the universe in which the Maya lived and thought. The chronological systems were their instrument of grasping the mysteries of *kinh* whose essence consisted of the divine countenances bearers of good and evil.

Thompson has rightly concentrated his attention on several Classic inscriptions on which are displayed likenesses of

4. For the description of the inscriptions and glyphs corresponding to these periods, see Thompson, *Maya Hieroglyphic Writing,* pp. 142–147.

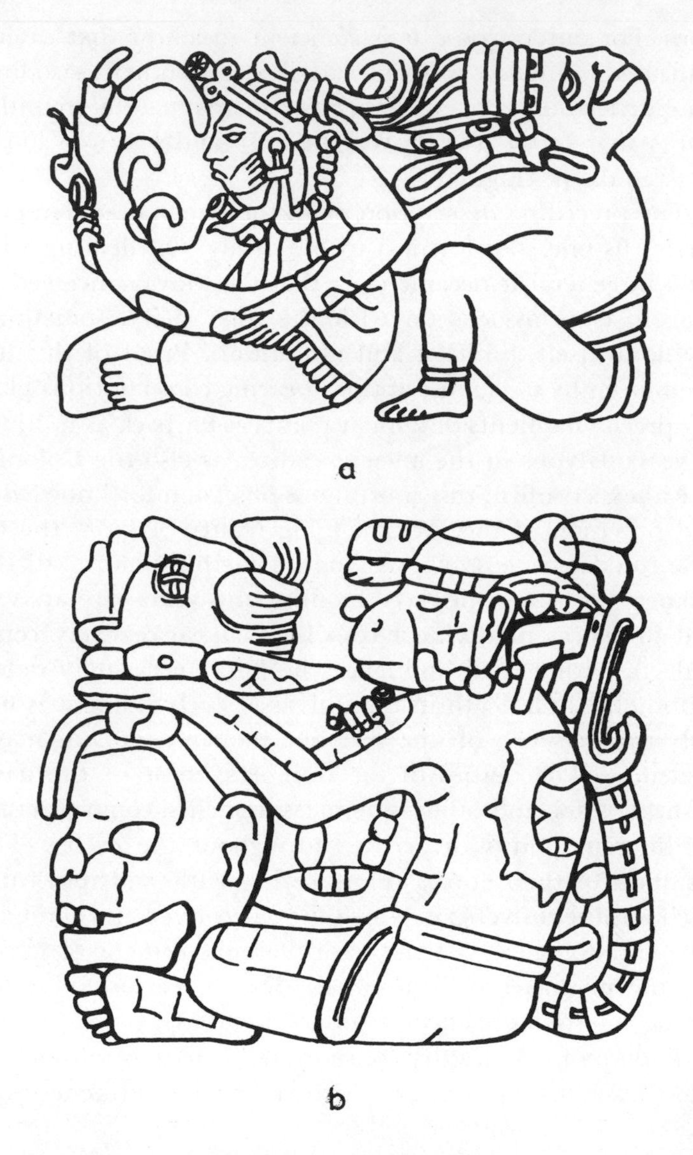

Figure 14. Divine bearers of time, after Thompson.
a) 0 *kins* (Copán D);
b) 16 *kins* (Yaxchilan, lintel 48)

gods, the personification of numbers, who, as in a never-ending relay race, bring with them the burdens of the days, months, and years. At the precise moment in which one of these periods comes to its completion, another deity takes up the burden and anew directs the flow of *kinh*. Examples cited by Thompson such as lintel 48 of Yaxchilan and stelae B and D of Quiriguá and D of Copán illustrate this (Figure 14).

In conceiving and measuring the reality of the different periods in terms of their completeness, Maya thought achieved a new form of expression through this image of gods as bearers of time. Their arrival at the end of a journey (*lub,* as termination or complete count in various Maya languages) is precisely the moment of the "weariness" of the gods. *Lub* also means "to become tired" in the various languages of this family. The new deities who in the same moment will take over the burden of time will carry it on their backs until, overcome by fatigue, they arrive at another place of rest which is the completion of one cycle and the beginning of another. Understanding thus the measures of time as repose-completion, one of the roots from which the idea of cycles is derived may be perceived. These are unending series of periods with moments that are at once ends and beginnings.

The Colonial Maya texts which speak of the twenty-year periods or wheels of the *katuns* confirm this peculiar conception of a universe in which the passage of time consists of arrivals, relays, and departures of divine forces. This is shown, for instance, in the "first prophetic wheel of a series of *katuns,*" published by Barrera Vásquez and reconstructed from various texts of the *Chilam Balam* books. In it we find the expression of the ancient symbols announcing the arrival of the diverse periods with the countenances and figures of the gods. Marking the renewal and entry of a burden of time, the text reads:

> 11-*Ahau* settles down with 13-*Ahau.*
> This is its word
> and the content of its burden shows it:
> Face of the heaven's birth
> is the seat of the 11-*Ahau Katun.*
> There its mat will be present.
> There its throne will be present.
> There it will reveal its word,
> There it will reveal its might.
> *Yaxal Chac,* Green rain,
> is the semblance of the *katun*
> that will rule in the sky. . . .
>
> BARRERA VÁSQUEZ, *op. cit.*:95–96

And after enumerating the fates of this period, in this case full of sorrows, the text describes the faces of the following *katuns.* Regarding the *katun* 9-*Ahau,* we are told that it is "the reign of *Ah Bolon Kin,* that of 9-Sun . . ." (*ibid.*:99). Of 7-*Ahau* it is proclaimed that "*Amayte Kauil,* deity of the four directions, will be its semblance in heaven" (*ibid.*:100). Of 12-*Ahau* it is affirmed, "here is that which manifests its burden . . . *Yaxal Chuen,* great-monkey-craftsman, is the face that it will have during its rule in heaven. There will be great masters, great sages, great magicians . . ." (*ibid.*:112).

Let us consider also one example of the survival of the ancient symbolism as related to the signs of the day-series in the prophecies of the *Chilam Balam of Kaua:*

> *Ix Kan*
> Lady of the Maize. Also wealthy,
> master in all the arts,
> *Ix kokobta,* the wryneck bird, is its omen.
> The precious singers, are its birds.
> *Chac Imix Che,*
> the red ceiba tree, is its tree.
> Sage.

Chicchan
Ah Tzal ti can, the serpent with the rattle
is its omen that comes with its tree.
Habin is its tree,
of fire is its soul.
Evil is its fortune.
Murderer.

Cimi
Ah cuy manab, the harbinger owl,
infamous the portent
that comes with its tree.
Murderer, equally evil is its destiny . . .

Lamat
A malformed dog is its omen.
Its face is that of a jaguar,
its hindquarters are those of a dog.
Meddlesome, gossiper . . .

Chuen
Countenance of a woodcarver,
craftsman of weaving is its omen,
master of all the arts.
Opulent all its life.
Good are all the things it will do. . . .

Ibid.:189–193

Expressions about the faces of gods with burdens of time—entering and being installed in order to perform their action during determined periods—continued to appear in the prophetic books for many years after the Conquest and occasionally up to relatively recent times. This was the result of the deep-rooted Maya concern with time.

What has been presented here about the best-known elements of Maya symbolism in the inscriptions, codices, and

Colonial texts, has introduced us to the unique complex of meanings inherent in the primordial reality of *kinh.* Upon this base some preliminary conclusions and some hypotheses can be formulated at this point:

a) The concept of time, an abstraction arrived at through experiencing the cyclic action of the sun and of the day which is its creature, was universally present among the Maya at least from the appearance of the first inscriptions of the Classic period. Proof of this is the ancient word *kinh,* having identical meanings in the diverse groups as well as the existence of its glyphic variants of which examples are found in codices of Postclassic times.

b) *Kinh,* sun-day-time, is a primary reality, divine and limitless. *Kinh* embraces all cycles and all the cosmic ages. That is why it is possible to make computations about remote moments hundreds of millions of years away from the present. Also, because of this, texts such as the *Popol Vuh* speak of the "suns" or ages, past and present.

c) The divine nature of *kinh* is not thought of as something abstract and shapeless. In it can be distinguished innumerable moments, each with its own face, carrying a burden which displays its attributes. Among the faces appearing in the diverse periods are those of the solar deity in all its forms and those of the gods and goddesses of rain, earth, corn, death, sacrifice, the great star, the moon, and hunting. These faces constitute the most significant nucleus of the Maya pantheon.

d) The time universe of the Maya is the ever-changing stage on which are felt the aggregate of presences and actions of the various divine forces which coincide in a given period. The Maya strove, by means of their computations, to foresee the nature of these presences and the resultant of their various influences at specified moments. Since *kinh* is essentially cyclic, it is most important to know the past in order to understand the present and predict the future.

e) The faces of time, mystical reality prompting the Mayan obsession, are the object of veneration. They determine and govern all activities. Thanks to them man knows the norms for agricultural labors, cycles of festivals, everything in life. The priests register the symbols and effigy of the time-gods as they arrive. They erect stelae, compose their books, and set the *katun* stones in place. Man sees his existence colored by time, presence, and action of all the countenances of the divinity.

Reflected in these conclusions lies possibly something of the essential relationship perceived by the Maya between time and the world of the gods. Thus one can also gain insights about some of the consequences touching on life and human acts. Nevertheless, to draw closer to the universe of *kinh,* it is necessary to seek new paths in our study. Among other things, one should ask, "What were the relationships of *kinh* with the visible and tangible world of spatial reality?"

Within the Maya view of a world distributed in four immense sectors—with several celestial levels, abode of the gods, and lower planes, the regions of the shades—what relations existed between time and cosmic space, also filled with symbols? Were time and space different aspects of the same primordial reality? If this should be so, should their thought be described as a peculiar type of pantheism which could be designated by a new term: *pan-chronotheism?* The application of such a concept undoubtedly might be arbitrary, being a sort of label devoid of significance in the case of Maya culture. It is therefore imperative to draw near to their own thought, seeking the meanings they gave to the spatial world in which they lived. Again turning essentially to the native sources and eliminating, as far as possible, ideas foreign to Maya mentality let us explore this theme: What was their conception of what we call space and reality within their philosophy of time?

CHAPTER IV

TIME AND SPACE

The image of a "spatial universe," as conceived by the Maya, has also left an imprint on the sources already known to us: inscriptions and symbolism in archaeological discoveries of the Classic and Postclassic periods; the three extant pre-Hispanic codices; the texts written by native authors and by Spanish chroniclers of post-Conquest times. The origin of these sources is varied: in respect to places of origin (highland or lowland groups), as well as in the diversity of epochs in which they were recorded or compiled. This situation helps to explain some variants that will be encountered in our study of the Maya image of the spatial universe. In spite of differing elements, however, occasionally betraying external influences or different patterns of evolution in some Maya groups, there is also a kind of common denominator in the symbolism and traits typical of what was the core of their vision of the spatial universe. In our approach to it, and before analyzing its relationships with the Maya concept of time, special consideration will be given to those fundamental elements which seem to be ever present throughout the evolution of this culture.[1]

Let us begin with the evidence available on what may be

1. It must be emphasized that the Maya spatial image of the universe had great similarities with the conceptions existing among the other cultures in Central America. Eduard Seler, among others, pointed this out in "Das Weltbild der alten Mexikaner" (Seler 1902–1923:IV, 3–37).

described as the great horizontal plane of the earth. The symbolism of the Classic period is the first of our sources. On stelae, monuments, and panels of not a few of the ancient ceremonial sites the earth symbol is shown either as the figure of a monster, fangs and claws of a crocodile, or with the form and head of fantastic saurians. Thusly, in conjunction with other cosmic elements, the representation of the earth appears in three of the most celebrated wall panels of Palenque: those of the Sun, the Cross, and the Foliated Cross. The earth monster, located in the lower level of the Sun panel, has two heads and is supoprted by as many jaguar deities, recalling in their function the *bacabs* of the Yucatec Maya, the deities which uphold the sky at the four quarters of the world. On the other hand, on that of the Cross the earth god appears with a monstrous mask on which rests the sacred tree that is the main theme of the panel (Figure 15). Finally, on that of the Foliated Cross,

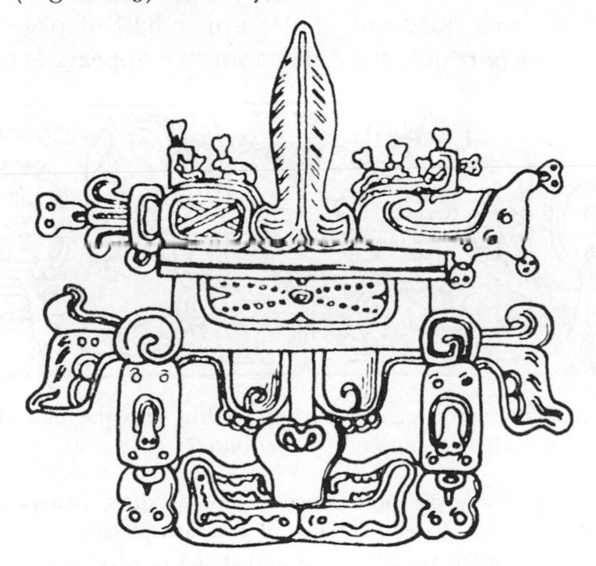

Figure 15. The monstrous mask of the earth, upon which the sacred tree rests. (*Panel of the Cross, Palenque*)

also serving as a base and support, its forehead bears the *kan* cross, symbol of jade and water. From this springs the maize plant.[2]

An earlier representation of the monster-deity of the earth is offered by Stela 1 from Tikal. On it, giving foundation to a world of symbols with cosmic connotations, the same deity is depicted with a prominent nose and accompanied by the sign of "zero" or completeness. Various other examples could be mentioned. Only one other, studied by Thompson (1960:12), will be recalled: that of Stela 7 of Yaxchilán. It exhibits the *Imix* calendric sign, which, as we have seen, refers precisely to the earth (Figure 16).

Reflecting beliefs that had continued into the Postclassic period, the codices also offer representations of the strange being upon which life prospers. Beginning with the *Dresden Codex,* two particularly expressive representations may be cited. Framed by calendar glyphs and by four deities situated on each quadrant of the upper half of page 3 (its only painted portion), the earth monster appears extended

Figure 16. A stylization of the earth crocodile with the sign of *Imix* as a headdress, after Thompson. (*Yaxchilán* 7)

2. A detailed analysis of the symbolism of these panels and of other masterworks at Palenque (such as the sarcophagus lid on which also appears the earth monster) is found in *La escultura de Palenque* (de la Fuente 1964:135–139). See also *La civilización de los antiguos mayas* (Ruz Lhuillier 1963:103–122).

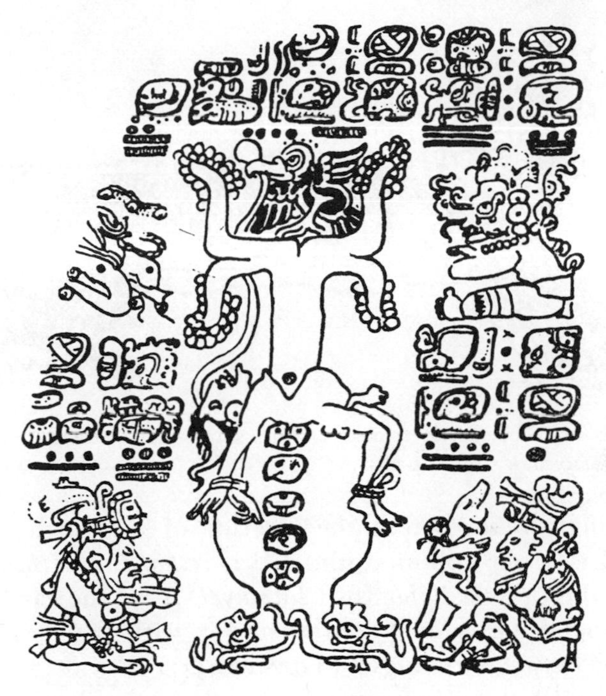

Figure 17. The two-headed earth monster. Above stands the *moan* bird. Four deities are portrayed at the extremes. (*Dresden Codex* 3)

with two heads and four paws which show what seem to be the scales of a reptile (Figure 17). At the highest point, over the hind part of the monster, is perched one of the cosmic birds. In the center, stretched over the monster's body is a human being in a position reminiscent of that of sacrificial victims. The harmony and symbolism of the composition seem to reflect one of the aspects of the earth within the context of the Maya vision of the universe. The other example, coming from pages 4–5b of the same codex, has already been pointed out by Tozzer. In the center of these two pages, with various divinities above, below, and to the right, stands the fantastic crocodile from whose fangs emerges the head of God D, a representation of the solar deity,

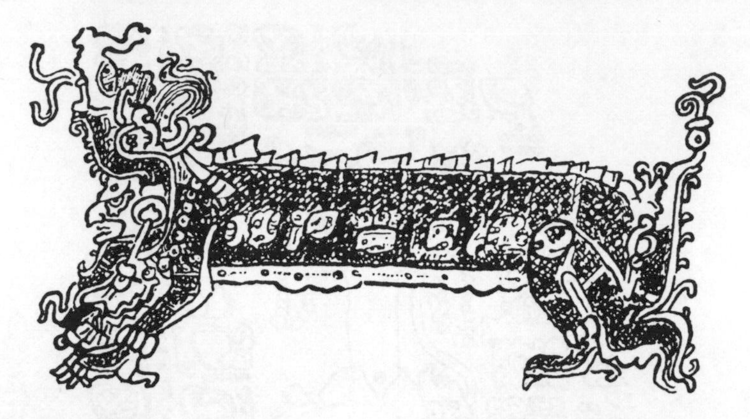

Figure 18. The earth monster. The solar deity emerges from its fangs. (*Dresden Codex* 4–5b)

seemingly reborn from the interior of the earth (Figure 18). In Tozzer's opinion, the god is *Itzam Cab Ain,* spoken of in the *Chilam Balam of Chumayel.* It is "the earth monster that appears in the Dresden 4–5 with the head of Itzamná between its fangs"[3] (Tozzer 1941:192 n. 1017).

It will not be imperative to examine in detail various other representations of the earth monster in this and the other two codices. We shall only cite the example on page 22 of the Paris manuscript, and on pages 16, 66, and 67 of the *Codex Tro-cortesiano* where the monster reappears in association with several deities and with other symbolic forms also connoting the Maya vision of the world.

Finally, as proof of the continuity of the concept of this deity as central symbol of the earth's great horizontal plane, we shall quote a fragment of "the prophecy of a 13-*Ahau*" that forms part of the Colonial texts of *Chilam Balam* published by Barrera Vásquez. There, once again the earth is expressly mentioned as being a crocodile-deity:

3. See also Tozzer and Glover, *Animal Figures in the Maya Codices* (1910:273–372).

The 13-*Ahau* is the moment
in which the sun and moon will come together and coincide.
It will be the night and at the same time the dawn
of *Oxlahún Tiku,*
13-Deity,
and of *Bolón Tiku,*
9-Deity.
It will be then when *Itzam Cab Ain,*
Wizard-of-water-earth-crocodile,
will create, will give birth to
enduring life on earth.

1948:146

Thus sources of the Classic and Postclassic periods, and even of the years following the Conquest, demonstrate the perdurance of this first element of the Maya vision of the spatial world.

To continue our study we will now analyze several complexes of symbols, attributes of the earth's surface in respect to its cosmic distribution toward the various directions. Now we shall proceed by reversing our order of research: first, consideration will be given to texts transcribed after the Conquest; afterward, the codices and inscriptions will be discussed for possible antecedents.

In the first pages of the *Popol Vuh,* the four-part distribution of the world is proclaimed and reiterated as something almost obvious:

> It is the original book,
> written in ancient times,
> but its face was hidden
> to him who seeks, the thinker.
> Great is its description, its account
> of how all was finally established,
> the heavens, the earth,

its four angles,
its four corners well marked,
the four of them well established,
their place chosen,
their measurements taken,
in the heavens, on the earth,
four angles,
four corners. . . .[4]

SCHULTZE JENA 1944:2

Rather than simply oriented to what we call the four cardinal points, the universe of the Maya appears distributed in four grand cosmic segments that in their turn converge on a point, the center, the fifth direction of the world. By relating the symbolism of the four segments with the concept of the earth monster, we find that the original figure of the great crocodile is transformed in turn into four beings, "each assigned to a world direction and each with its distinguishing features . . ." (Thompson 1950:12).

Among the Yucatec Maya, the four monsters, both terrestrial and celestial, occasionally received the name of *Itzamnás.* They are addressed thus in one of the spells from the text known as the *Ritual of the Bacabs,* published by R. L. Roys in 1965. The incantation which cures one from obstruction in the respiratory tract, insistently invokes the four Itzamnás, "crocodiles of the house,"[5] in relation to the four regions and colors of the universe:

Crocodile of the red quadrant,
come to me,
thirteen are the waters

4. Translation based on the Quiché text from the Schultze Jena's edition of the *Popol Vuh,* fol. 1 v (1944:2).

5. The term *Itzamna,* translated at times as "dew from the heavens" (*Itz en caan*), may be also interpreted, as Roys notes, as "crocodile of the house" (*Iyzam-na*) (Roys 1965:152–153).

of my red gutter,
when I guard my rear
behind the east sky.

Crocodile of the white quadrant,
come to me,
thirteen are the waters
of my white gutter,
when I guard my rear
behind the north sky.

Crocodile of the black quadrant,
come to me,
thirteen are the waters
of my black gutter,
when I guard my rear
behind the west sky.

Crocodile of the yellow quadrant,
come to me,
thirteen are the waters
of my yellow gutter,
when I guard my rear
behind the south sky.

ROYS 1965:64

The incantation continues with typical monotony. Through all of it the four *Itzamnas* are related with the colors, the waters, the winds, and other symbols of the four "measurements" (*Kaan*) or quadrants of the world. One last text, selected from various which could be cited, of the *Chilam Balam of Chumayel,* offers other elements of the rich complex inherent in the four-part distribution of the earth. Following the same scheme, east-north-west-south, there now appear the sacred stones, the birds, the seeds, the cosmic ceibas, and the beings corresponding to each quadrant of the world:

The red flint stone
is the stone of the red *Mucencab.*
The red ceiba tree of abundance
is his arbor
which is set in the east.
The red bullet-tree is their tree.
The red zapote . . .
The red-vine . . .
Reddish are their yellow turkeys.
Red toasted corn is their corn.

The white flint stone
is their stone in the north.
The white ceiba tree of abundance
is the arbor of the white *Mucencab.*
White-breasted are their turkeys.
White Lima-beans are their Lima-beans.
White corn is their corn.

The black flint stone
is their stone in the west.
The black ceiba tree of abundance
is their arbor.
Black speckled corn is their corn.
Black tipped camotes are their camotes.
Black wild pigeons are their turkeys.
Black *akab-chan* is their green corn.
Black beans are their beans.
Black Lima-beans are their Lima-beans.

The yellow flint stone
is the stone of the south.

The ceiba tree of abundance,
the yellow ceiba tree of abundance,
is their arbor.
The yellow bullet-tree is their tree.
Colored like the yellow bullet-tree are their camotes.
Colored like the yellow bullet-tree
are the wild pigeons which are their turkeys.
Yellow green corn is their green corn.
Yellow-backed are their beans. . . .

Chilam Balam of Chumayel 1933:64

These relatively late texts concerning the cosmic regions preserve an extraordinarily rich symbolism: the earth monster appears now in the four sectors of the world. Each region is tinted with its own color: red in the east, white in the north, black in the west, yellow in the south, and green in the center. In each sector grows the primeval ceiba together with its corresponding cosmic bird. As recorded in the Yucatec texts, there reside in the four regions the *pahuatún* (gods of the wind), the *chac* (Lords of rain), the *balam* (protectors of the fields), and the *bacab* (the supporters of the heavens). Other deities also maintain a close relationship with the earth. Among them are the young maize god, the jaguar, the god of death, and others who are also associated with different temporal periods.

It is now time to examine the possible antecedents of this symbolism in the most ancient sources: the codices and carved inscriptions. First of all it must be observed that, just as we encountered in these sources multiple representations of the earth monster, we will now find evidence of a universe distributed in four great sectors which originate in a central point or region. A good illustration of this is given on pages 75–76 of the *Madrid Codex* (Figure 19). In the center appears a cosmic tree with two divinities seated in its shade, facing several glyphs of *ik,* symbol for the wind and life. All is surrounded by the twenty day-signs. This is the central region of the earth from which spring the four great quadrants, indicated by strips of footprints. Each sector displays its glyph: south to the left, north to the right, west above, and east below. And it should be emphasized that these glyphs, with the sole exception of the one for the north, are exactly the same used to denote the cosmic directions in the earlier carved inscriptions of the Classic period. Such is the case of various monuments at Tikal, Piedras Negras, Naranjo, Copán, Quiriguá, and Palenque. As in the post-Conquest Maya texts, each sector represented

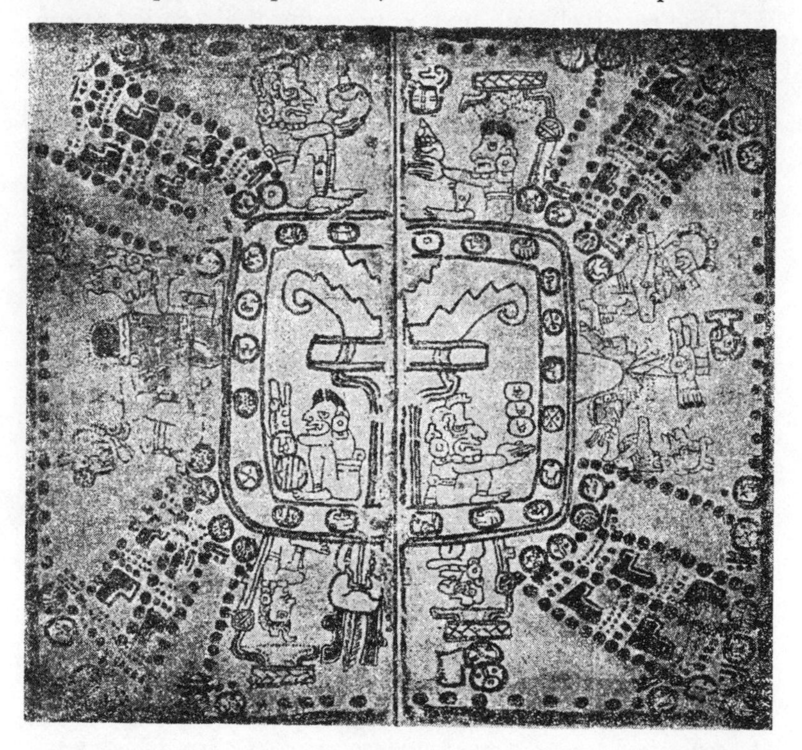

Figure 19. The fourfold division of the world with directional and cosmic symbols. The area in the center is surrounded by the glyphs of the days. (*Madrid Codex* 75–76)

in the *Madrid Codex* also has its own deities, painted with their corresponding colors.

Several other examples from the codices can also be cited which clearly reflect these and other aspects of the Maya concept of the fourfold distribution of the world. Three of these will be cited. On page 31a of the *Madrid Codex* appears *Chac,* the rain god, situated among four smaller figures, stylized frogs (*uoob*), with glyphs corresponding to the four cosmic regions (Figure 20). On pages 25c–28c of the *Dresden Codex,* beginning with the glyph for the east, are shown the four *acantun,* stones or pillars that, according to Landa, were erected in the ceremonies at the beginning of a new year. The pillars represented in the codex take on the

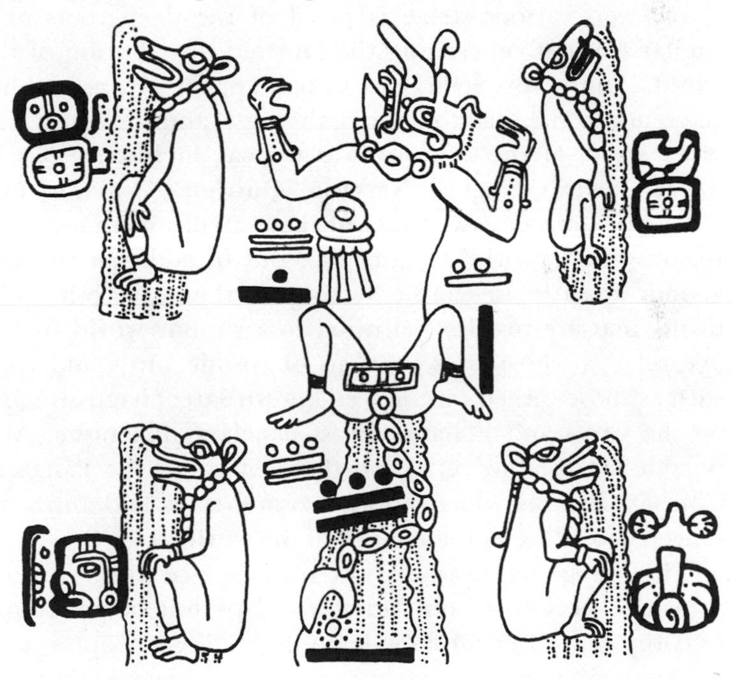

Figure 20. In the center: *Chac,* the god of rain. In the four extremes appear the *uoob,* frogs, companions of *Chac,* with the glyphs of the cosmic directions. (*Madrid Codex* 31a)

form of trees with the signs of *cauac* and *che,* which seem to indicate their nature: they are cosmic trees related to rain and to the world of celestial forces. Also represented there are the four cosmic birds as sacrifice offerings in the hands of each of the four divinities who preside over their corresponding regions.

One last example is found on pages 30c–31c of the same *Dresden Codex* which exhibit four representations of God B above as many cosmic trees with the signs and colors proper to the east (red), the north (white), the west (black), and the south (yellow).

In reference to the carved inscriptions of the Classic period, the presence of the already-mentioned "directional glyphs" on various stelae is proof of the deep roots of a similar concept concerning the fourfold distribution of the world. Thusly, on Stela A of Copán appear glyphs making reference to the east, the north, the west, and the south. On other stelae of the same site—as well as in inscriptions of Tikal, Piedras Negras, Naranjo, Quiriguá, Nakum, and Palenque—one finds variants of these glyphs connoting the regions of the world[6] (Figure 21). But in addition to these glyphic records, in Classic symbolism there are other elements that are manifest allusion to a similar world image. We refer to the representations of cosmic birds and trees such as those carved on the well-known sarcophagus lid and on the Cross and Foliated Cross panels at Palenque. Also notable are the low reliefs on the Temple of the Panels at Chichén Itza on which appear cosmic trees and birds arranged according to the regions of the world.

The foregoing examination of evidence from diverse areas and periods of the Maya world permit us an insight into the more outstanding features of these peoples' con-

6. Among the stelae portraying "directional glyphs," are Tikal 12, Piedras Negras L 12, Naranjo 24, Copán A, E, J, and P, Nakum D, Quiriguá M, and Palenque, Inscription M.

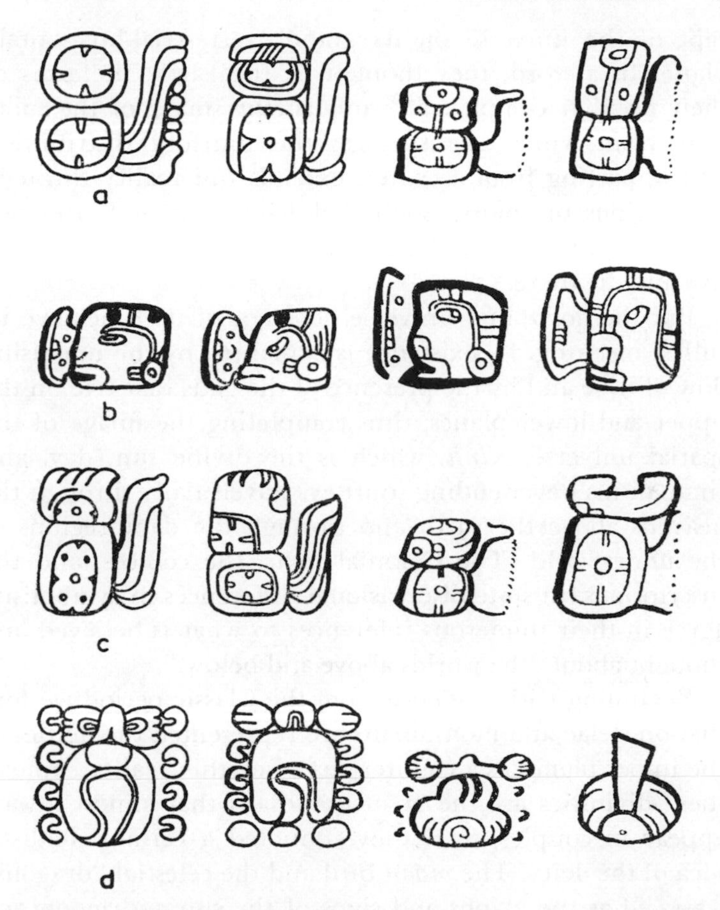

Figure 21. Directional glyphs (orientation of the years) in the carved inscriptions and codices.

a) East: Nakun D; Palenque Inscription M; *Dresden Codex* 43c; *Madrid Codex* 50a.

b) North: *Dresden Codex* 44c; *idem* 31c; *Madrid Codex* 25d; *idem* 50a.

c) West: Palenque Inscription M; Naranjo 24; *Dresden Codex* 31c; *Madrid Codex* 50a.

d) South: Copán T; Quiriguá M; *Dresden Codex* 31c; *Madrid Codex* 25d.

(*Source:* THOMPSON, *Maya Hieroglyphic Writing*)

cept of the universe on its grand "terrestrial-horizontal" plane. In a word, they thought of the earth, in terms of their peculiar complex of symbols: the image of the monsters from which life springs; the fourfold distribution which, parting from a center, extends full reality throughout regions of colors, populated with gods and primeval forces, cosmic birds and trees, and also human beings that live and die there.

But the horizontal universe does not of itself achieve its fullest meaning. Its existence is explained by the unceasing flow of time and by the presence of the gods that live on the upper and lower planes, thus completing the image of the spatial universe. *Kinh,* which is the divine sun, day, and time, in his never-ending journey, travels daily through the heavens above the earth and through the dark regions of the underworld. The Colonial texts, the codices, and the inscriptions, in spite of occasional differences in symbolism, agree in their numerous references to what is believed and thought about "the worlds above and below."

Beginning with evidence from the Classic period we find that on stelae and monuments are represented the deities of the upper planes as a counterpart of earthly reality. Among these divinities are the great ophidians that almost always appear in couples, as if they connoted a certain dualistic idea of the deity. The *moan* bird and the celestial "dragons" —as well as the glyphs and signs of the sun and moon and sometimes of the great star (Venus)—also form the plastic image of the world above. The truncated pyramids, with superimposed bodies and a sanctuary on the highest level, may be, as some investigators have maintained, another reflection of the heavenly floors on which the gods live.

In Maya architecture, since Classic times, there has existed another eloquent form of plastic representation of the heavenly beings. Here we refer to the elements called "façades of Quetzalcóatl" (Venus) by Eduard Seler. On them appears a great mask with the features of a monster or an ophidian, whose mouth functions as the entrance to the temple. In connection with this, Thompson has remarked that on several of these façades one also encounters often celestial glyphs, not only that of the planet Venus (Thompson 1939:391–400). This is the case of the façade of Temple 22 of Copán, of the House or Temple of the Dwarf at Uxmal, and of the eastern annex of the Nunnery at Chichén Itza. These and other façades, as well as some low reliefs on altars and temples exhibiting a similar monstrous mask associated with celestial symbols, portray the dwellers of the upper planes.

Finally, the presence in the inscriptions of the thirteen variants of god-faced numbers seems to show, as has been noted, the thirteen divine beings that exercise their influence in as many levels in the upperworld.[7] A confirmation of this is probably found in the glyph of the *moan* bird accompanied by the symbol of heaven and the number thirteen in numerous classic inscriptions and in various places in the *Dresden Codex* (Figure 22).

An evocation of the lower worlds underneath the earth monsters, are the glyphs of the nine gods of the night and of the regions of darkness. Among these glyphs are the ones that represent the color black, the shell, the eyes of the death god, and, above all, the countenances of the deities or of a group of them with the sign of a hand grasping a head with simian features. This hand is preceded by the number 9, and is found in stelae inscriptions such as those of Piedras Negras 25 and 36 and in others that could be cited. Following Morley, we may also recall the vestiges of nine anthropomorphic sculptures in the left chamber of Temple 40 at Yaxchilán. These also seem to represent the

7. According to Thompson, these are the thirteen gods who can almost certainly be identified with the thirteen gods of the days (Thompson 1960:210).

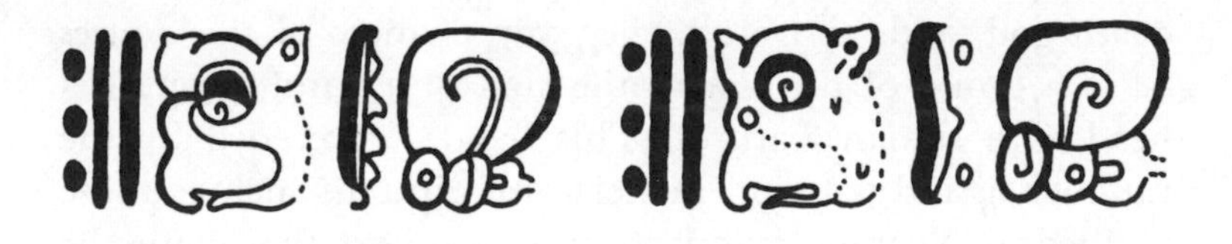

Figure 22. The *moan* bird in symbolism and in the codices.
a) The bird inclined over planetary symbols in Copán H;
b) Glyphs of the *moan* bird. *Dresden Codex* 16c, and
c) *idem* 18b.
(*Source:* THOMPSON, *Maya Hieroglyphic Writing*)

gods of the nine levels of the underworld (Morley 1947: 233).[8]

In the codices we find also abundant and precise evidence related to said image of the universe, already shared by the Classic Maya and other peoples of Middle American culture.[9] The glyphs of the nine gods of the underworld appear in the three Maya codices although the faces of the celestial deities appear only in the *Dresden Codex*. Apart from the glyphs, the codices bear numerous representations with the symbols of ophidians, the *moan* bird, the sun, the moon, the great star, and others belonging to the upper levels. There we also encounter the signs of the God of Death, the shell, the color black, and the "night sun," all pertaining to the lower world. Moreover, on pages such as 74 of the *Dresden* (Figure 23) or 31b of the *Madrid* some of the cosmic cataclysms seem to be depicted; they are possibly the endings of "ages" or "suns" on which, as mentioned in the *Chilam Balam of Chumayel,* the various planes of the universe went through an upheaval.

On dealing with the prophecies of a *Katún* date, *11-Ahau,* in the *Chilam Balam of Chumayel* again we read about the central elements of the ancient image of the world and of the symbolism already familiar to us. In this case these elements are incorporated in a dramatic description of the struggle between the thirteen celestial deities and the nine gods of the underworld. The burden of the *Katún* has provoked a cataclysm involving the three planes of the universe. At the end, the primigenial ceibas are re-

8. Morley adds that "along the eastern base of the square tower of the Palace at Palenque there were found the badly destroyed remains of nine similar figures, also in stucco, which probably had represented these same nine gods" (Morley 1947:233).

9. Regarding the parallel concepts of the Nahua of the Central Highlands, see *Aztec Thought and Culture* (León-Portilla 1963:48–61).

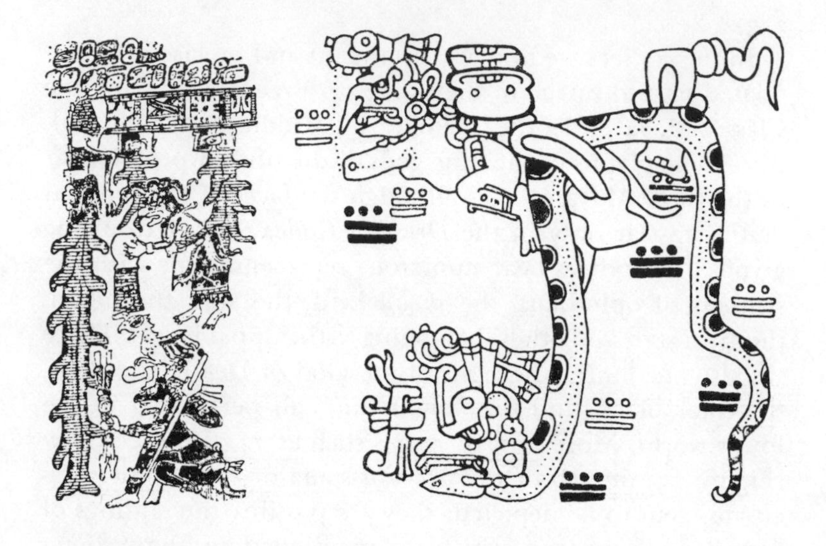

Figure 23:

a) Probable representation of a cosmic cataclysm *Dresden Codex* 74;

b) *Chac,* god of rain, related to a cosmic serpent, *Madrid Codex* 31b

born in the five regions of the earth as a sign and memory of the violent encounter of divine forces that destroy and then re-create the universe through the various "suns," which are also the great cycles of time:

> The *11-Ahau katun*
> is when *Ah mucen cab* came forth
> to blindfold the eyes
> of *Oxlahún ti ku,*
> Thirteen-deity . . .
> With their countenances blindfolded,
> dawn ended for them
> and they knew not what was to come.

When *Oxlahún ti ku,*
thirteen-deity,
was seized
through the deeds of *Bolón ti ku,*
Nine-deity;
then will descend ropes and fire,
and stone and stick,
and there will be smiting with stone and stick
when *Oxlahún ti ku,*
Thirteen-deity,
will be seized.
Then
his head shall be cracked,
and his face will be slapped,
and he shall be spat upon,
and carried upon another's shoulders,
and stripped of his insignia,
and covered with soot.
And the quetzal bird, and the green *yaxum,*
will be ground and eaten,
along with his heart . . . [of *Oxlahún ti ku*],
through the deeds of *Yax bolón dzacab,*
Great-nine-begetter.

He will take hold
of the thirteenth floor of the heavens,
and will scatter the dust
which will fall off the seeds,
and the point of the shelled ear of corn,
the bone of corn,
here on the earth,
the place of his heart,
because *Oxlahún ti ku,*
Thirteen-deity,
does not respect the heart of man's sustenance . . .

The heavens will collapse,
and so will the earth collapse,
when the *katun* reaches its end . . .

Then will rise *Cantul ti ku,*
Four-deity;
the four *Bacabs,* outpourers,
who will devastate the earth.

At the end of this devastation,
will rise *Chac Imix che,*
primeval red ceiba,
the column of the heavens,
sign of the dawning of the world,
tree of the *Bacab,* outpourer,
on which will perch *Kan xib yuyum,*
the yellow bird.

Then will also rise
Sac Imix che,
The white ceiba to the north;
there will perch *Sac chic,*
the white bird.
Support of the heavens,
sign of annihilation
will be the white ceiba.

Then will also rise
Ek Imix che,
primeval black ceiba,
to the west of the flat land.
Sign of annihilation
will be the black ceiba.
There will perch *Ek tan picdzoy,*
bird of the black breast.

Then also will rise
Kan Imix che,
primeval yellow ceiba,
to the south of the flat land,
as a sign of annihilation.
There will perch *Kan tan picdzoy,*
bird of the yellow breast . . .

Then also will arise
Yaax Imix che,
primeval green ceiba,
in the central region,
as a sign and memory of annihilation.
It is the one that holds the plate and the
drinking vessel,
the mat and the throne of the *katuns*
that live by means of it.

BARRERA VÁSQUEZ 1948:153–155

The above Colonial text confirms the deep roots of the symbols under which the Maya image of the universe was integrated throughout the various stages of the cultural evolution of these ancient Americans. The words cited from the *Chilam Balam* introduce us, furthermore, to the problem of the relationships between time and space. On describing the struggle that convulsed all the levels and sectors of the universe, it is expressly stated that this catastrophe was the burden of an 11-*Ahau katun* date. In this we have a first token, though late, of the significance of time in the ambit of spatial reality. But following the method we have adopted, it will be necessary to seek in more ancient sources for indications or proofs regarding the patterns of relationship between space and time.

As a first step, the fourfold division of the earth's surface will be considered. Inscriptions on the stelae are again our oldest source. As demonstrated by several Maya scholars, the glyphs of the cosmic directions appear on stelae, occasionally accompanied by year signs or by signs of the day on which a year terminates. "Taken in conjunction with the world-direction glyph, it reads 'To the east (north, west, or south) the count of the year' " (Thompson 1960:251). In other words, the years that travel on the backs of their corresponding divine bearers, the four signs designated in Yucatec Maya as *Kan, Muluc, Ix, Cauac,* are linked to the four world regions and to the complexes of symbols inherent to each area. Thus, for example, there are signs "of the year to the east" on stelae P of Copán and D of Nakum. Glyphs "of the year to the west" appear on Copán E and Naranjo 24, and of the year "to the south" on Copán E.

The codices and later texts confirm this ancient manner of conceiving time as something continuously permeating space. Thanks to color symbolism—red signifying east; white, north; black, west; yellow, south; green, center—it is often possible to identify in the codices the regions toward which specified time cycles are oriented. It is true that there are alternates and changes in the symbolism. It is clear, nevertheless, that, in the minds of those who elaborated the codices, time and its deities penetrate and tinge with their attributes the different sectors of the spatial world which is also the abode of the gods.

Proof of this close relationship between time and space is given, among other examples, by the above-cited pages 30c–31c of the *Dresden Codex*. On these, along with calendric glyphs accompanying four representations of the God B, he of the long nose, with the same number of cosmic trees, there appear also the signs and colors proper to the four world regions. The above-cited pages, 75–76 of the *Madrid Codex,* also acquire a deeper significance if we observe that in the representation of the four great quadrants, with their direction glyphs, the whole of horizontal space appears surrounded in its turn by the twenty day-signs. Thus is reiterated the constant time-space relationship (see Figure 19).

In reference to the years that journey on the backs of their divine bearers, and are successively oriented toward each of the spatial regions (a concept first appearing on the stelae) we are informed by Father Diego de Landa that the same idea persisted throughout later epochs. Making use, as might be expected, of Yucatec Maya year names and using the Christian calendric expression "dominical letters" as equivalent of "year-bearers," the friar writes:

> The first then of these dominical letters is *Kan*. The year which this letter stood for was the omen of the *Bacab,* whose other names are *Hobnil, Kanal Bacab, Kan Pauah Tun, Kan Xib Chac*. They assigned this one to the direction of the South. The second letter is *Muluc*. They assigned him to the East; his year was the omen of the *Bacab,* which they called *Can tizic nal, Chacal Bacab, Chac Pauah Tun, Chac Xib Chac*. The third letter is *Ix*. His year was the omen year of the *Bacab,* which they call *Sac cimi, Sacal Bacab, Sac Pauah Tun, Sac Xib Chac;* they assigned him to the direction of the North. The fourth letter is *Cauac*. His year was the omen of the *Bacab,* which they call *Hosan ek, Ekel Bacab, Ek Pauah Tun, Ek Xib Chac,* they assigned him to the direction of the West.[10]
>
> LANDA 1941:136–138

Various late texts from Yucatán again confirm the persistence of these ideas about the orientation of time, partic-

10. In the above enumeration of the deities corresponding to the cardinal points, the color of the different years precedes the names of the beings known to the Yucatec Maya as *Pahuatuns,* who dwelt in the underworlds. The same occurs in the cases of the *bacabs* or supporters of the world, and in that of the *chacs* who resided in the clouds and were known as *chacs* of several colors.

ularly of what we call "spatialization" of the years. In the Maní manuscript we find the following example:

> 13-*Ahau: Ah Pulá* died.
> There remained six years
> in order to complete
> the count of 13-*Ahau,*
> the year that was counted
> to the east.
> On 4-*Kan*
> the month *Pop* fell to the east. . . .
>
> BARRERA VÁSQUEZ, *op. cit.*:65

What could be considered an exposition of the system of correlation between time and space, a sort of commentary to the "wheels of the *katuns,*" is included in the late *Ixil Text,* gathered by the Maya scholar Juan Pío Pérez. Among other things it offers a discussion of "the seats and burdens of the years" (*ah cuch haab*): the *Kan* years toward *Lakin* (the east); the *Muluc* years toward *Xaman* (the north); the *Ix* years toward *Chikin* (the west); the *Cauac* years toward *Nohol* (the south). The commentary reads:

> When [the year] has gone round
> the four quadrants,
> east, north, west, south,
> it is said to be 1-*Katun.*
> The same thing is said
> when 1-*Cauac* begins,
> and each of the years goes through this. . . .
>
> SOLÍS ALCALÁ 1949:340–341

In a succession that only concludes at the end of a cycle in order to give birth to a new one, the burden of the years becomes manifest in each quadrant of the world. This is true not only of the years but of the months, days, and all god-faced periods of time. On penetrating and acting within the spatial universe, all these temporal units communicate to it their being and attributes, thus joining the other divine forces which preside over the diverse sectors. Fundamentally, the countenances of time are the same as those of the deities of the earth's surface: if *Imix* is the first of the days, it also evokes the primeval reality of the world.

Knowledge about the socio-religious and political structure of modern Mayan lowland communities offers a new example confirming the universal incidence of this essential time-space relationship. As pointed out by chroniclers and anthropologists, the communities and villages were fundamentally distributed in four districts or principal divisions, each one oriented toward its corresponding sector of the universe. Thus, "a similar division in four districts existed also in Chichén Itzá as well as in Itzamkanac, capital of the province of Acalan, inhabited by the Chontal Maya and situated in the Southeast zone of the peninsula . . ." (Villa Rojas 1961:32). The archaeologist Michael D. Coe has recently shown on analyzing the rites that pertained to the last five days in the year (the *Uayeb* of the Yucatec Maya), that it is precisely the yearbearer's orientation which determines even now the order of transmission of political and religious powers among the chiefs of the districts or divisions in the community. "The idea of the city divided into quarters is shared, to be sure. . . . It is the principles that underlie such quadripartitions that vary and which must be analyzed . . ." (Coe 1965:109). "What makes the Maya community model unique, however, is that they alone seem to have hit upon a permutating time count as a kind of automatic device to circulate power among kin groups of the primitive state . . ." (Coe 1965:112). Due to the time count or what is the equivalent, the orientation of the years, the four districts or basic divisions of the community, being a reflection of the spatial universe, successively provide chiefs to govern in the age-old order of east-north-west-south.

Kinh, the universe of deified time, appears thus spatialized in multiple forms through its relationships with the great terrestrial plane and its quadrants of colors.

Turning now to the upper and lower strata which complete the spatial image, we will see that in both the alternating presence of *kinh*—the sun, maker of the day and of all the cycles of time—is equally manifest. The face of the old god, the deity with the flat nose and the canine tooth projecting from the corner of his mouth, is portrayed in association with bands or stripes of astronomic or celestial glyphs pointing at his presence and actions in the upper strata of the universe. This is to be found in the symbolism of numerous monuments dating back to the Classic period. The same *kinh* is often represented as if it were about to be devoured by the fangs of the fantastic crocodile that repeatedly connotes the earth. At other times *kinh* is shown as being reborn from the same fangs after having accomplished his journey through the somber regions of the underworld. On the lower levels, *kinh* is both nocturnal sun and time. His symbols then become those of the jaguar, the dog, the god of death, and of several other beings.

Precisely due to the cycles of *kinh,* to his successive presences above and below, the reality of the three world planes achieves its fullest unity and meaning. As noted by Spinden, the flat-nosed god, the effigy characteristic of *kinh,* "is a universal sky divinity with powers extending over the day as well as the night" (1957:69). A good confirmation of this is given, among other examples, by a representation in the *Dresden Codex,* 56a. There, the aged solar countenance appears under a band of astronomic glyphs in the center of an almost rectangular figure, half black, half white. Said figure from the *Dresden* is probably a symbol of an eclipse though

Figure 24. The face of *kinh* in the table of eclipses of the *Dresden Codex* 56a

it also calls to mind the alternating of day and night in the four regions of the world and in the entire universe (Figure 24).

If *kinh,* sun-time-deity, is present on the celestial levels, on the earth, and on the underworld planes, it should not surprise us to find that the gods of the numbers, days, months, years, and other cycles (being different bearers of the burdens of time), also appear in the symbolism of the upper and lower worlds. An example of this is probably offered by the bas-reliefs of stucco on the walls of the Tomb of the Inscriptions at Palenque. On these walls are carved nine personages carrying "in one hand the so-called mannequin-sceptre crowned with what probably represents the rain god; they bear in the other hand a circular shield with traits characteristic of the solar god" (Figure 25) (Ruz 1954: 89). Because of their being situated in the Tomb of the Inscriptions, the nine personages may, perhaps, be related to the nine gods of the nether regions. Regarding, now, the upperworlds, representations as those of the glyph of the

Figure 25. Two of the nine personages on the walls of the crypt of the sarcophagus in the Temple of the Inscriptions at Palenque, after Ruz

moan bird with the sign of heaven and the number 13, in the already mentioned Classic inscriptions, seem in turn to point at the complex that later came to be known as *Oxlahún-ti-ku,* the thirteen celestial deities. A thorough analysis of the symbolism of the Classic monuments and later codices certainly proves that the deities associated with glyphs of celestial connotation, as well as those related to the underworld, are the same as those which appear in various forms in the hieroglyphs of the diverse counts of time. Among these deities stand out the sun in all its variants, the moon and the great star; the gods of rain and of corn; the serpents, crocodiles and other cosmic monsters; and the Lord of death. In short, the most significant members of the old Maya pantheon are present. On the upper and lower planes and on the terrestrial surface, furthermore, these gods grow richer with new attributes which bring their beings to completion. These are the gods who dynamically color and permeate all of the spatial universe of the Mayas.

The text we have cited about the upheaval of all the world's levels due to the burden of an 11-*Ahau katun,* acquires from this point of view its most complete significance: the spatial universe exists, changes, dies, and is reborn in each of the "suns" or ages as a consequence of the actions of the gods or countenances of time. Space is not static. It is the complement, the framework of colors, which from moment to moment sets up the stage for *kinh.* Upon it, as if determined by the rules of a game, or as in a drama developing in cycles, time displays its diverse countenances and masks successively. It is thus that *kinh* gives life, destroys and re-creates without end the reality in which men move and think.

Does this mean that space and time basically constitute a homogeneous entity? Or, which is the same, were space and time, more than being merely related, a perfect identity in the core of Maya thought? So far, present analysis does not permit the formulation of a definitive reply. What can be affirmed, however, is that if space exists as the handiwork of the gods and has in itself divine meanings, all the deities present and acting in space are the changing countenances of time. Isolated from time, space becomes inconceivable. In the absence of time-cycles, there is no life, nothing happens, not even death. The colored regions, divorced from *kinh,* sun-day-time, would become utter darkness devoid of all meaning. The world of the gods would be a mere absence, and the flight of the *katuns* would mark the end of reality. There would be a return to primeval darkness, without the cosmic ceibas, without the sun, the moon, the great star, without human beings, without any meaning whatsoever.

Time, on the contrary, is the life and origin of all things. That is why its study and measuring is the supreme concern of the wise men the *ah Kinob* ("those of the solar and temporal cults") as they were designated by the Maya of Yucatán. It is to one of these *ah Kinob,* named *Napuctun,* that is attributed a text in the book *Chilam Balam of Chumayel* which underscores the deep relationship between time and space. The words of *Napuctun* are certainly an echo of the ancient knowledge of the Maya concerning the significance of time as the beginning and root of that which exists. The following is his account of what happened when for the first time a twenty-day period was integrated and the monthly cycle was born.

> Thus it was recorded
> by the first sage . . .
> The first prophet, *Napuctun,*
> the priest, the first priest.
> This is a song of how the *uinal*
> the twenty-day cycle,
> came to be created
> before the creation of the world.
> Then he began to march
> by his own effort alone. . . .

Then said his maternal grandmother,
then said his maternal aunt,
then said his paternal grandmother,
then said his sister-in-law:
"What shall we say when we see man on the road?"
These were their words
as they marched along,
when there was no man as yet.
Then they arrived there in the east
and began to speak.
"Who has passed here?
Here are footprints.
Measure it off with your foot."
So spoke the mistress of the world.
Then he measured the footstep
of our Lord, God the Father.
This was the reason it was called
counting off the whole earth.
This was the count,
after it had been created by the day 13-*Oc,*
after his feet were joined evenly,
after they had departed there in the east.

Then he spoke its name,
when the day had no name,
after he had marched
along with his maternal grandmother,
his maternal aunt,
his paternal grandmother and his sister-in-law.
The *uinal,*
twenty-day cycle,
was created,
the day, as it was called,
was created,
heaven and earth were created,

the stairway of water,
the earth, rocks and trees;
the things of the sea
and the things of the land
were created.

On 1-*Chen* he raised himself to his divinity,
after he had made heaven and earth.
On 2-*Eb* he made the first stairway.
It descended from the midst of the heavens,
in the midst of the water,
when there were neither earth,
rocks nor trees.
On 3-*Ben* he made all things,
as many as there are,
the things of the heavens,
the things of the sea,
and the things of the earth.
On 4-*Ix* sky and earth were tilted.
On 5-*Men* he made everything.
On 6-*Cib* the first candle was made;
it became light,
when there was neither sun nor moon.
On 7-*Caban* honey was first created,
when we had none.
On 8-*Edznab* his hand and foot
were firmly set,
then he picked up small things on the ground.
On 9-*Cauac* the lower world was first considered.
On 10-*Ahau* wicked men went to the lower world
because of God the Father,
that they might not be noticed.
On 11-*Imix* rocks and trees were formed;
this he did within the day.
On 12-*Ik* the breath of life was created.

The reason it was called *Ik*
was because there was no death in it.
On 13-*Akbal* he took water
and watered the ground.
Then he shaped it
and it became man.
On 1-*Kan* he first created anger
because of the evil he had created.
On 2-*Chicchán* occurred the discovery
of whatever evil he saw
within the town.
On 3-*Cimí* he invented death;
it happened that our Lord God
invented the first death.
On 5-*Lamat* he established
the seven great waters of the sea.
On 6-*Muluc* all valleys were submerged,
when the world was not yet created.
Then occurred the invention
of the word of our Lord God,
when there was no word in heaven,
when there were neither rocks nor trees. . . .

The *uinal* [the twenty-day cycle]
was created,
the earth was created;
sky, earth, trees and rocks
were set in order;
all things were created
by our Lord, God the Father.
Thus he was there in his divinity,
in the clouds, alone and by his own effort,
when he created the entire world,
when he moved in the heavens in his divinity.
Thus he ruled in his great power.

Every day is set in order
according to the count,
beginning in the east,
as it is arranged. . . .

ROYS, *Chilam Balam of Chumayel:*116–119

The words attributed to *Napuctun* poetically reiterate and clarify the relationship: the heavenly planes, the earth and the nether worlds stand due to the birth of the month and the days. Life and reality exist thanks to the toil of *kinh,* sun, day, and time deified.

So it is that the theme of the spatial omnipresence of time and its divine character reappears insistently as a constant from the early carved inscriptions of the Classic sites, through the codices, and the late Colonial writings all the way to the traditions and beliefs of some of the modern Maya, as ethnographic research has shown. Upon this basis the moment has arrived for us to attempt a synthesis of the integral image in which Maya astrological endeavors appear conjugated with the world of meanings time held for the priest and wise man. In our essay to reconstruct said image, we seek an answer to the original question: What did the universe of time mean to these ancient Americans within the complexity of their life and cultural evolution?

CHAPTER V

MAN IN THE UNIVERSE OF *KINH*

Priests and wise men of the Maya Classic period were neither the first nor the only ones in Central America to be concerned with the subject of time. It is well known that in earlier ages, in lands not remote from what was to become the Maya world, there throve peoples in possession of different calendar systems. Evidence of this is given by the inscriptions on the stelae of the *danzantes* from the early level of Monte Albán I in Oaxaca. They indicate that those living there employed the 260-day count some centuries before the birth of Christ. This calendrical computation was to be known as *tzolkin* among the Maya; *tonalpohualli* among the Nahua; and *pije* among the Zapotec (see Caso 1946:I, 113–145 and 1965:931–947).

From the coastal lands along the Gulf of Mexico, probably the cradle of Central American mother culture, proceed also ancient inscriptions related to the counting of time. In them, most probably before any other people in the world, the Olmec assigned a value to numbers in function of their relative positions within their computes. Some examples are the celebrated Stela C of Tres Zapotes with what may be a date equivalent to 31 B.C., and the Tuxtla statuette registering a date corresponding to A.D. 162. From the culture area of Izapa comes also the carved inscription of Stela I (or *Herrera*), found in El Baúl, Guatemala. Said inscription seems to antedate by 256 years the oldest known

Maya monument with calendar computations, *i.e.*, Stela 29 of Tikal, dating from A.D. 292 (see Coe 1957:597–611; Coe 1965a:756–763).

These discoveries clearly show that the endeavor in computing temporal cycles originated and was diffused in Central America prior to the flourishing of Classic Maya culture. Proof that the Maya were not the only people to possess calendar and chronological knowledge is afforded us by parallel evidence left by some of their neighbors and contemporaries. These were groups of the Central Highlands, of the area of Oaxaca, of Veracruz and of the Pacific coast, in short, all the peoples of high culture of Central America. But although the Maya were not the first nor unique in concentrating on the theme of time, they were certainly the most obsessed with it. They developed what probably was a common Central American heritage to the point of creating new chronological systems with units of measurement functional enough to achieve the highest degree of precision in the reckoning of time.

Maya obsession with time becomes evident from the beginning of the Classic period, at least since the third century A.D. Full knowledge, however, of such calendar systems as the Initial Series, which soon was diffused throughout the entire Maya lowlands, presupposes a long evolution. From this it may be inferred that the pre-Classic inhabitants of these lands, like those of the Gulf Coast and Oaxaca, somehow participated in the invention of the most ancient calendar systems or at least were among the first to incorporate them in their cultures. The fact is that the beginnings of the Maya Classic period coincide with the carving of stelae and monuments on which are preserved the first indications of the extraordinary development later to be achieved in the computation of time. It is also notable (and worth repeating at this point) that Maya chronological strivings persisted to such an extent that they have been rightly considered a "trait or key pattern" in their cultural evolution (Vogt 1964:35). For this reason to inquire into the deepest meaning that time and its computations held for the Maya may be equivalent to probing into what was the soul of their culture.

Various are the ways of approaching the meanings which the concept of time had in the thought and symbolism of the ancient wise men. Our point of departure has been that of seeking the possible existence of a concept whose contents might refer, somehow, to what was the primordial reality of time to the Maya. If this concept existed, above all in the thought of the wise men, evidence of it would remain in the language itself, in the glyphs, the symbolism, and the texts.

Our research, referring not only to the Maya of late epochs but also to those of the Classic period, first began in the field of linguistics. Recent comparative studies of the different languages belonging to the Maya family, show, among other things, that there is a word that, having cognates or related terms in all these tongues, obviously was a part of the vocabulary used in the ancient proto-language. This word, already familiar to us, is *kinh,* whose connotations of sun-day-time are constant in all the languages of this family.

The correlation of the word *kinh* with the complex of inscriptions and symbolism, beginning with the Classic period, has permitted us, apart from verifying the frequent reiteration of this concept, to perceive something of the richness of its semantic content. Hieroglyphs of *kinh,* as much in the inscriptions of calendric content as in the nonchronological texts or clauses, are among the most frequent. This has been demonstrated by Zimmermann and Thompson in their ample catalogues of glyphs. The persistence of several of the variant *kinh* glyphs in the codices and symbolism of the Postclassic period, as well as many allusions in the Colonial texts, prove a basic continuity, in spite of secondary differences, of this ancient form of conceptualizing time.

Thus attending to the evidence from the different epochs marking the evolution of this culture, it was possible to amplify the study of what constitutes the nucleus of meanings inherent to this fundamental concept. *Kinh,* the sun, is at once the aged countenance of a primordial deity: it is he who makes the day and the heat, and then penetrates the levels of the underworld to reappear in the east and ascend anew to the celestial regions. The sum of his cycles, of his untiring coming and going through all the planes of the world, is precisely the essence of the periods of time. The twenty-day cycles or months, the years, the *katuns,* and the whole of the periods, being themselves nothing but time and a consequence of the journeys of the sun, also participate as he does in the nature and attributes of divinity. Hieroglyphs and symbolism give proof of this: all periods of time are not only presided over by their corresponding deities but they themselves are divine personifications. Also the numerals accompanying the glyphs of the various cycles display countenances of gods because they are the carriers of time.

The temporal universe, created again and again by the ancient face of the solar deity, makes possible through its cycles the arrival of all the other god-periods. These, with their own rhythm and measure, bring with them diverse forms of burdens, good and bad fates, with which all reality is tinted and permeated in a succession that never terminates. The deities of the numbers, the days, the months, the *katuns* and *baktuns* make their entrances and are acting until their journeys are done. Then they pass their burdens to the other time-gods as they arrive. The cycles of *kinh,* time that flows without ceasing, will make possible the return of the distinct god-periods. Due to this, the wise men, in order to know the future, reckoned the positions and recurrences of the gods in moments of the past, some of them distant by hundreds of millions of years.

Past cosmic ages, the "suns," are precisely the great days of *kinh*. Through them were reborn the earth, a deity with a monster's figure, the world of the primeval ceibas, the universe of colors with its celestial floors and the abode of the dead. Through the action of *kinh* all becomes present in time. Its burdens color the four segments of the world. The countenances of the god-periods are successively oriented toward the great quadrants, determining the destiny and life of humanity and all existing things. Space, and that contained therein, acquire their true meaning due to the cycles of *kinh*.

Thus did Maya man come to conceive his universe from this most peculiar point of view of the cosmic atmosphere of *kinh*. But, if he was almost hypnotized by time, his attitude toward it was never passive. In close connection with what could be called his *chronovision* of the universe, he developed his computations and calendar systems. He discovered new units of measurement making possible adjustments and corrections with admirable precision. So it was that his own religious world view was born and developed in a somewhat different way from that of the other Central American cultures.

Although it is true that the Maya pantheon, beliefs, rites, and ceremonies present obvious similarities with parallel institutions belonging to peoples of the Central Plateau, the Gulf Coast, or the region of Oaxaca, a closer examination reveals significant differences. Since the Classic period the Maya, like other Central Americans, worshiped deities whose attributes were rain, maize, wind. They adored others associated with the earth, and with the celestial realities, the sun, the moon, the great star, and also with the lowerworlds, the region of somber levels where the dead dwell. Concepts also paralleled in neighboring groups were those of the cosmic quadrants, the regions of the different colors, the primeval trees, as well as the ages or "suns" through which the universe had existed. Finally, many of the rites and ceremonies, ruled by the two best-known calendar forms (the 260-day count and the solar year), were likewise a common possession revealing the Maya share in an ancient cultural heritage.

Beyond these and other similarities that could be mentioned, there existed, however, among the Maya—especially during the Classic period—attitudes and trends that determined significantly different forms of thought. Their wise men did not simply relate the gods to moments of time falling under their special protection and in which it was necessary to celebrate rites and sacrifices in their honor. The sages conceived of time itself as the primordial reality, the deity of multiple countenances, periods, cycles, which in alternating journeys and with the possibility of returns in a never-ending flow, communicates his burdens to all the planes and quarters of the world. It was, therefore, *kinh,* the most obvious of the time-countenances, who acquired supreme rank.

Extraordinary plastic representations of this conception are offered by several stelae carved during the Classic period in Yaxchilán. As noted by Spinden on describing the upper portion of Stela 10 (1957:69–76, and Plate LI), on it we have an image of the celestial reality: in the central position, above the cross of *kan* (water-jade) and other celestial symbols, there appears the old lord, *kinh,* time that governs all. On either side, in symmetrical squares, there are two deities, probably the sun and the moon, each having a scepter crowned with a mask, which also portrays the sun. Below the symmetrical squares, and as if united to the frame of the celestial symbols, on either side are seen three faces, all of them variants of the same figure with the solar eye and the prominent eye-tooth. The faces on the extreme ends protrude from the stylized jaws of the well-known terrestrial monster (Figure 26).

Contemplated in its entirety, the bas-relief displays four sectors; the two symmetrical squares above, and the two groups of faces below. *Kinh* is in the center, resting on the band of celestial signs which, descending like a frame on both sides, impress a harmonious unity on the image of the

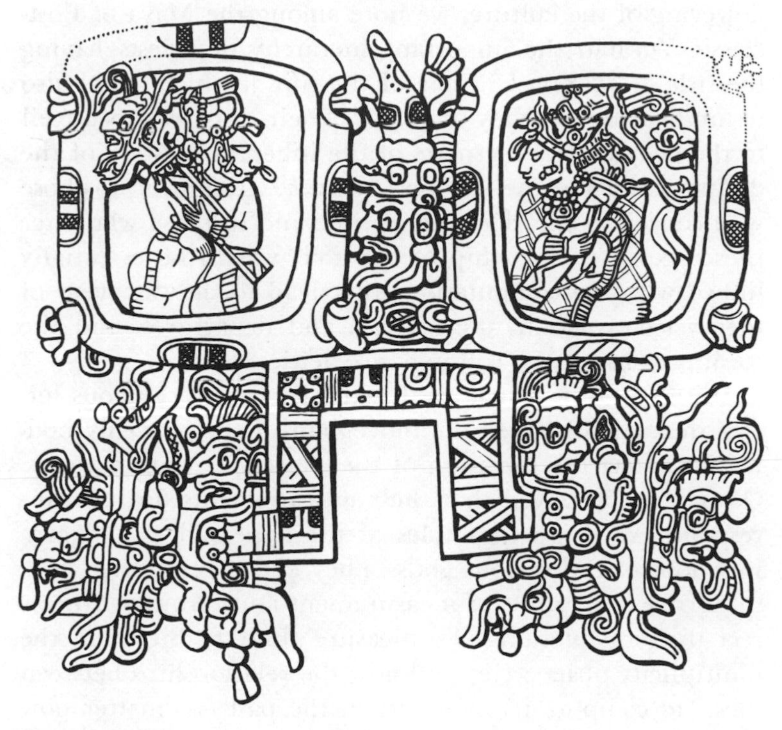

Figure 26. The universe of *kinh*. In the center appears the face of the god with the solar eye and prominent eye-tooth. In the lateral squares are two deities (probably the sun and moon) with scepters ending in masks of *kinh*. Below, and also on both sides, as if emerging from the strip of celestial symbols, are two groups of three faces bearing the same portrayal of *kinh* (upper part of Stela 10, Yaxchilán.

upper world. Presiding above, he creates the cycle of the days: he appears in the heavens; then, on one side, he enters the fangs of the monster, journeys through the regions below, finally reappearing, on the other side, from the same monster's fangs. The universe of the faces and masks of *kinh* thus achieves its most perfect expression.

As a persistent vestige of the primeval significance of this ancient conception, originated in the period of greatest flowering of the culture, we note among the Maya of Postclassic Yucatán the important hierarchy of priests having the title of *ah Kinob* ("they of the sun"), which could also be understood as "they of time." Their duty was to unveil to the people certain aspects of the inherent mystery of the divine periods in the universe of *kinh*.[1] According to those who conceived this doctrine of time and to those who later preserved it, the worship of the gods was to be essentially linked with the computations that lead to the discovery of the cycles of action, the arrivals and departures, and the conjunctions, of the countenance-deities.

With a zeal in which are fused science and religious fervor, the sages observed the movement of the heavenly bodies to perfect the precision of their calendar computations. Often they perceived that their measurements did not correspond exactly to the cycles of the appearances and conjunctions of the various gods. They endeavored, therefore, to discover new units of measurement and formulae to correct their calculations. To measure time, to hit upon the multiplicity of its cycles, to know the relationships between these, to compute its moments in the past (no matter how

1. It is significant that the title of *ah Kinob,* originally referred to that hierarchy of Maya priests, was to be applied, much later, during the Colonial days, to the preachers of the Christian gospel, that is, to those who presented themselves as bearers of a new divine message (see Bernardo de Lizana, *Historia de Yucatán,* cited by Tozzer 1941, 27, footnote 148).

remote) became the supreme form of wisdom, the one which drew man to the mystery of divinity.

The consequences of these efforts, at least in part, are already known to us. Within religious thought arose a most peculiar conception of a mythology interwoven with calendrical computation. Maya knowledge about time could be called a pure science if from it were eliminated its absorbing religious connotation and what we today would describe as its astrological applications. But as it existed historically, in its intrinsic relationship with a universe conceived as a manifestation of *kinh* and of all his countenances, it crystallized in what perhaps might be designated as a unique form of *mathematized* religion and mythology.

This is confirmed by the hieroglyphs expressing calendar computes. Not only *kinh* (time), but also the signs and symbols of the days, months, years, *katuns*—all the distinct periods, and also the numbers themselves—all were endowed with divine faces and attributes. Inversely, the deities' very being, their moments of action, the periodicity of their cycles, their orientation in space, their fates and conjunctions are in their turn measurements of time. They are ciphers accompanied by symbols, mathematical units which are to be computed in order to propitiate and worship the integral reality of the universe of *kinh*.

For this reason, among the forms of worship during the Classic period particularly significant was the erection of stelae and the recording upon them and on other monuments of the precise computations that were also measurements of the journeys of the divine bearers of time. The same can be said about the *katun* stones of the Postclassic epoch. In chronicle III of the *Memorial of the Katuns of the Itzá* we have a reminder of this. There, record is preserved of the setting up of these stones.

12-*Ahau,* the *katun* stone was set up in Otzmal,
10-*Ahau,* the *katun* stone was set up in Sisal,

8-*Ahau,* the *katun* stone was set up in Kancabá,
6-*Ahau,* the *katun* stone was set up in Hunancthi,
4-*Ahau,* the *katun* stone was set up in Atikuk,
2-*Ahau,* the *katun* stone was set up in Chacalná. . . .

BARRERA VÁSQUEZ 1948:72

And referring to the same form of cult, Fray Diego López de Cogolludo writes:

And when these periods reached five, which compose twenty years, they called it *katun.* Then they place a carved stone on another carved one, fixed with quicklime and sand in the walls of their temples and houses of the priests. This can be seen today in the edifices that have been mentioned and in some ancient walls in our monastery at Mérida. . . .

1954, I:337

Fray Diego de Landa, on explaining the order and form in which the Yucatec Maya venerated these "markers" of the *katuns,* notes something that corroborates the idea of the deity-periods that move, passing their burdens from one to another:

The order which they used in counting their affairs and in making their divinations, by means of this computation, was this—they had in the temple two idols dedicated to two of these characters. They worshipped and offered homage and sacrifices to the first, according to the count from the cross on the circle shown above [see Figure 4 in this book], as a remedy for the calamities of their twenty years. But for the ten years, which remained of the twenty of the first idol, they did not do anything for him more than to burn incense to him and to show him respect. When the twenty years of the first idol had passed, he began to be succeeded by the destinies of the second and (they began) to offer him sacrifices,

> and having taken away that first idol, they put another in its place, in order to worship that for ten more years. . . .
>
> Had there been no Spaniards, they would have worshipped the idol of *Buluc Ahau* until the year 1551, which would have made ten years; and the tenth year (1551) they would have put in place another idol to *Bolon Ahau,* and would have rendered homage to him, the prognostics of *Buluc Ahau* being followed till the year 1561; and then they would have taken him from the temple and would have placed there the idol *Uuc Ahau,* and the prognostics of *Bolon Ahau* would have been followed, during ten years more, and thus they gave their turn to all. In this way they worshipped these *katuns* of theirs during twenty years, and during ten years they were ruled by their superstitions and their frauds. . . .
>
> LANDA 1941:168

Since in the Postclassic period the Initial Series system had fallen into disuse, the cult associated with monuments commemorating moments in time referred to the different *Ahau,* the solar countenances on which each of the *katuns* concluded. In the case of the Classic stelae, on the other hand, on recording longer periods, the Maya venerated the conjunction of several countenances at a determinate point in the Long Count or grand system of reference to the universe of *kinh.* Thompson's reading of the chronologic inscription of Stela D at Copán is pertinent here. The priests and wise men, on recording on them the result of their computations, left in their glyphs something that could be read in the following manner:

> The count of the *tuns* [the years]. The moon goddess is the patroness. The *Chicchan* god bears the burden of the *baktun.* The earth god of the realm of the dead bears the *katun.* The earth god carries the weight of the *tun.* The deity who symbolizes completion doffs the burden of the *uinal,* and has

run his course with the *kin*. The god of the dead rests from carrying the day *Ahau,* and our divine youth of the maize likewise has reached the end of his stage with *Ch'en* upon his back. THOMPSON 1950:153

The recognition and adoration implied by the recording of the cycles have yet another meaning that should be underlined. The chronological inscriptions were also in themselves great chapters in which were described the history of a universe whose essence is time. Through them the wise men related the memory of the most significant moments in the perennial becoming of *kinh* within the spatial world. Their knowledge of time, tinged as it was with religiosity, myths, and astrological applications, meant also a profound effort to understand the ultimate reality, omnipresent and cyclical, which determines the events of a history that is at once cosmic and divine.

From this point of view, it will be easier to understand the wise men's insistence in emphasizing the intrinsic relationship between the burdens of certain moment-deities and the consequences for human beings derived from them. Thus, as an example we will mention one from the first chronicle in the *Chilam Balam of Chumayel* already pointed out by Roys. It refers to the destiny proper to the *katun 8-Ahau* that always carried with it, for a millennium, the force that obliged the Itzá to change their place of residence:

Katun 8-Ahau recurred approximately every 256 years, and for a thousand years every time a *katun* of this name occurred, the Itzá were driven from their homes, no matter where they were living at the time. Late in the seventh century A.D., they were expelled from Chichén Itzá after their first occupation of that city. In the middle of the ninth century they were driven out of Chakanputún. At the end of the twelfth century

Plate III. Glyphs of complete figures: "deities who are bearers of time." Stela D at Copán. The reading is: 9 *baktuns,* 15 *katuns,* 5 *tuns,* 0 *uinals,* 0 *kins;* 10 *Ahau,* 8 *Ch'en.* (*From* THOMPSON, *Maya Hieroglyphic Writing*)

> they were again driven from Chichén Itzá by Hunac Ceel. About the middle of the fifteenth century, Mayapan was sacked and destroyed; and strangely enough it was again in a *katun 8-Ahau,* at the end of the seventeenth century, that the Spaniards conquered the last Itzá stronghold at Tayasal, which was the end of this remarkable nation.
>
> ROYS, *Chilam Balam of Chumayel* 1933:136, note 3

Cases such as this, exhibiting the relationship between the destinies of the *katuns* and, generally, between the burdens of the different periods and their consequences in the life of man, must have corroborated the conception of time as a universal determinant. The ancient concern in finding through the calendar the norms of the agricultural cycles and the moments to propitiate the gods, thus attains a fuller meaning. To ascertain the units of measurement of *kinh,* to know the order of its alternations, to investigate its past burdens, leads to the prediction of its future recurrences. Thusly, religious thought, supported by observations and calculations, unites its peculiar form of cosmic history with an astrological knowledge so vitally important throughout the life of the Maya.

The feasts and ceremonies, according to available evidence from the Postclassic period, were often occasions on which the *ah Kinob,* priests of the sun and time, made public their prophecies. For instance, during the feast of the month *Uo,* as described by Landa, the priests:

> . . . took out their books and spread them out on the fresh boughs which they had for this purpose, and invoking with prayers and devotions an idol named *Kinich Ahau Itzamana,* who they say was the first priest, they offered him gifts and presents. . . .
>
> This having been done, the most learned of the priests opened a book and looked at the prognostics of that year,

Plate IV. Stucco glyphs with the dates 9 *baktuns,* 14 *katuns,* which form part of the inscription of an initial series. Originally from Palenque (*National Museum of Anthropology, Mexico*)

and he manifested them to those who were present. And he preached to them a little, recommending to them the remedies for their ills. . . .

LANDA, 1941:153–154

Those who recognized the presences of the different moment-deities would find "remedies for their ills." They would be able to seek, thanks to the rites and sacrifices, and with the aid of the computations, the favorable destinies, those which confronted with adverse fates, would neutralize contrary influences. In this way it was possible to escape absolute fatalism and open the door to knowledge leading man to better acting and thinking at prescribed moments.

Plate V. Stucco glyph with the date 3 *Pop*. It forms part of the inscription of an initial series. Originally from Palenque (*National Museum of Anthropology, Mexico*)

The discovery of these relationships between the moments and periods of *kinh,* and human destiny, corresponded to the priests. They were the ones who indicated the days favorable for ceremonies such as the giving of a name to a child, the admission of a youth to a school, the celebration of a wedding, the beginning of a war, the consecration of a ruler; the healing of a patient, the days for selling and buying and those for sowing and reaping.

But though the possibilities of astrological prediction and the measuring of agricultural cycles seemingly explain the origin of the Maya preoccupation with time, to reduce the thinking of the sages to this alone would not be doing them justice. With their astrological preoccupations coexisted the knowledge which, through the symbols, had led to the discovery of a universe conceived from the point of view of an ultimate and all-embracing reality: time. The relationship with the ancient myths was preserved, and that which was belief and need for prediction became fused with rigorous measurements and computations. So was born a most unusual religio-mathematical vision of the universe, the fruit of highly refined and precise minds.

It is certain that the universe of the Maya was populated with countenances of gods, which were the forces acting in the quadrants of colors and in the celestial and lower regions. But differing totally from any form of animism, Maya thought had discovered the measurements of the cycles which with intrinsical order rule whatever happens in the universe. The divine forces were neither indeterminate nor obscure; their action can be foreseen by means of observations and computations. As in no other culture, the priests and wise men made out of time computations formulae for ritual and worship. In the inscriptions they commemorated with mathematical rigor the moments in which the action of the god-periods had left their imprint in the world. The elaboration of eclipse tables in their codices permitted them, for example, to dialogue with the gods, ordering the ceremonies and manners of acting in terms of what was going to happen because they already knew what the divine sequence of the universe was to be.

The entire life of the Maya thus presented itself oriented by a cultural pattern manifested in their institutions essentially related to the theme of time. Thus the religious cult prospered and with it symbolism, art, and their unique kind of science—in a word, life and the great and small actions of every day. Obsession with time, therefore, came to be a unifying factor in this culture.

If one would insist in comparing the peculiar focus of the Maya world view—though knowing that comparisons can falsify the object of the study—it would be necessary to have recourse to a conception, different and remote, but which at least presents a certain degree of affinity. We refer to what seems to be the central thought of the *Tao Te Ching*, the celebrated Book of Tao attributed to the Chinese sage Lao Tze.[2] We are told there about the path and norm which is the primordial reality and origin of all things. Tao is also the cosmic atmosphere that pervades and governs everything. Man will never be able to comprehend such a reality, but he ought, nonetheless, to seek it and, above all, bring his life into harmony with it. Tao is the rhythm of all that happens in the world. Truth and justice consist of seeing, even though faintly, and of following the path marked by Tao.

The above brief recollection of this ancient Chinese concept may show some sort of similarity with the Maya idea

2. Among the various direct translations from the Chinese of the *Tao Te Ching,* see that prepared by R. B. Blakney. *The Way of Life, Lao Tzu, Wisdom of Ancient China,* New York, The New American Library, 1955. A detailed commentary about the concept of *Tao* is offered by Holmes Welch, *Taoism: The Parting of the Way,* Boston, Beacon Press, 1965.

of *kinh,* the deified time that likewise permeates the universe and through its cycles governs all events. As in the case of the followers of Tao, the Maya also strove to approach the mystery of *kinh.* Its alternations and sequences having been discovered, man could properly abide in the world and adapt his life to *kinh's* rhythm and influences.

But if there could be similarities between such distant forms of thought, the differences are many. One should be emphasized in particular. Maya speculations regarding *kinh* had their origin in observation and computations. The priests knew what had been and what was to come through calculations in which with mathematical precision they made adjustments to set their measures of time in accord with the changing reality of the universe. Thus if for a moment comparison seemed to help, soon what is different

Plate VI. Complete figure: a period bearer. Originally from the "Palace" at Palenque (*National Museum of Anthropology, Mexico*)

becomes manifest—in our case, that which was exclusive of Maya thought. To penetrate more deeply into the meaning of this world view, rather than comparing it with alien concepts, it is necessary to attend to the already analyzed elements derived from sources which date back to the different periods of this culture.

Drawing from the evidence we have gathered and discussed throughout this work, let us attempt a final and comprehensive résumé of our theme.

Kinh is a cosmic atmosphere with visages of gods who become manifest cycle after cycle. From the beginnings of the Classic period the sages discovered units of measurement which encompassed the great ages through which the world and humanity have existed. Space and time were inseparable. The spatial universe was an immense stage on which the divine faces and forces were oriented, coming and going in an unbroken order. The Maya sages had found the key to extricate meanings and to foresee the future. The norm of life was to attune with what were and would be the burdens of time. Since the latter were computed in terms of the completion of the different periods, once their entire reality was known, it became possible to predict their returns and recurrences. The arrivals of the time-bearing deities could be determined beforehand. Thus it was feasible to situate oneself under the influence of those whose burdens were favorable. The wise men and priests, therefore, being those who knew the calendrical intricacies, were the ones who showed the people what suited them best.

Time signified this and much more for Maya consciousness. To compute its measurements was to create unusual forms of cosmic historiography and mathematical mythology. Since the cycles were gods, knowledge of time was the root of theological thought. The symbols of temporal connotation, the date-deities, communicated their meaning to everything that was produced in history. Art, for instance, could be described as that which expressed the tense harmony of a universe where the masks and figures of gods, each with its corresponding date-glyphs, occupied the principal place because they denoted presences and absences which affected mankind. Human figures whose representation proliferates in the art of sites such as Palenque, Yaxchilán, and Piedras Negras, and even in ceramics such as the celebrated vases of Chamá or the Petén, achieved their fullest meaning within the same stage where were measured the arrivals and departures of the divine visages, the bearers of time. If other peoples managed to forge different visions of the world, windows from which to look out in order to understand their universe, the Maya invented a cosmovision that, because it was history, measurement and prediction of the total reality, whose essence was time, it could more appropriately be named *chronovision*.

In the Maya consciousness of time were joined and reconciled (as one could know the sequences of the cycles), the universe of the gods and the world of the colored directions where men lived. In adverse, and even in fatal moments, the *chronovision* of the wise men always permitted the discovery of meanings. Perhaps because of this, with the hope of recovering the ancient meaning of existence or finding a new one in its stead, some Maya groups surviving the Conquest continued or remade as best they could the wheels of the *katuns* and the books of the prophecies. Clinging to the theme of time in order to save themselves, they also bequeathed to the world a last testimony of the ancient *chronovision* which, with all its variants, was the soul of a culture that lived for almost two thousand years.

Our original question about the reasons for the chronological obsession of the Maya receives at least the beginnings of an answer. The Maya art and science of measuring time, probably born before the beginnings of the Classic period, extraordinarily elaborated through the latter and with diminished strength even in modern isolated communities, integrated a culture pattern, the basis of many other institutions. In this pattern the Maya found norms for everyday life, for astrological knowledge, for the order of the feasts with their rites and sacrifices, for their economy, agriculture, and commerce, for their social and political systems. But above all their passion for time came to produce their *chronovision,* the conception of a universe in which space, living things and mankind derive their reality from the ever-changing atmosphere of *kinh.*

As no other people in history, the Maya perceived and lived the mysteries posed by a universe whose deepest substratum is time. For centuries they cultivated their obsession. Theirs was the desire of knowledge but also a concern for salvation, an attempt to discover the supreme order of things. Thus they conceived their myths, they created symbols, used the zero, invented new systems to adjust and correct their computations. They became worshipers of the primordial reality, omnipresent and limitless. To harmonize with that reality was the most precious aim in life. The wisdom of their priests and sages led them to attempt to discover their place on earth, and also to spy on the mysteries of the divine rhythms of the universe.

Agreeing strongly with the words of Eric Thompson, we must recognize that "no other people in history has had such an absorbing interest for time as the Maya and no other culture has ever developed in similar form a philosophy to abridge so unusual a theme . . ." (Thompson 1954: 316). To pass over the *chronovision* of the Maya would be to deprive this culture of its soul.

APPENDIX

THE CONCEPTS OF SPACE AND TIME AMONG THE CONTEMPORARY MAYA

BY

ALFONSO VILLA ROJAS

The ancient Maya image of the universe, according to the original and scholarly interpretation presented here by Dr. León-Portilla, offers the opportunity of perceiving in its essential traits the vast complexity of that world of ideas, a framework for the sages and priests of that part of Middle America. Their conception of time as something divine and eternally flowing, without beginning or end, distributed in recurrent cycles saturated with "burdens" or determinants of the destiny of man and the universe was united with their religious system and with their peculiar manner of conceiving the spatial structure of the universe. This constitutes one of the most elaborated creations of the human mind in its eternal desire to penetrate the secrets of existence.

A question may be asked: To what point have persisted, in their basic lines, those ideas that were so deeply rooted in the pre-Hispanic patterns of life? To what degree do those concepts of time and space persist in the contemporary Maya world? To a great extent the reply to these questions can be found through the rich and varied ethnographic material now available concerning native groups scattered within the vast territory inhabited by the modern Maya. This holds true especially in parts of Southeastern Mexico and Northwestern Guatemala.

We consider that this effort to trace survivals may bring us a better understanding of certain pre-Colonial concepts. It may also lead us to a more satisfactory comprehension of the acculturative process through which said Maya groups have passed. It should be noted that this is no easy task. Four centuries of contact with ways of life, imported with and after the Conquest, have given rise to hybrid, though coherent, cultural systems that are neither an exact replica of the pre-Hispanic pattern nor even less an exact copy of the European one. What exists today is a *sui generis* product constituted by elements of both origins that have managed to accommodate themselves in a functional whole.

Taking all this into consideration, it is clear that there do not exist today uses or customs of Maya origin completely identical to those practiced in pre-Columbian antiquity. Work in the cornfields differs on many points from the way it was carried out then. Today metal tools are used; furthermore, many of the old rites and beliefs accompanying agriculture have disappeared. The *Chacs,* benevolent spirits in charge of rainfall for the fields, no longer are subject to *Hunab-ku,* the supreme god of the ancient Maya, but to the "True God" introduced by the Christian missionaries. A similar statement could be made about other cultural complexes derived from the ancient forms of native life. To varying degrees, all have suffered transformation in order to adapt to the new cultural situation. To keep these circumstances in mind will serve as a cautionary measure for the interpretation of data to be presented.

On the other hand, in spite of the aforementioned adaptative process, one can still recognize not a few ancient traits in uses and customs that have persisted among the more isolated groups within the Maya area. Pre-Columbian survivals are specially obvious in respect to material culture but also are noticeable in concepts which even today are present in their peculiar image of the world. A good example of this is afforded by recent studies carried out among Maya groups in the state of Chiapas, Mexico, by a team of anthropologists headed by Dr. Evon Z. Vogt of Harvard University. The results of these studies have appeared in the book *Zinacantan: A Maya Community in the Highlands of Chiapas* (Vogt, 1969). This ethnologist's description of the indigenous religious situation in that region is illuminating:

> All Mayas today consider themselves Catholics (except for insignificant minorities here and there which have joined Protestant churches), but this certainly does not mean that their Catholicism has obliterated aboriginal cosmological ideas. We have found a number of ancient Maya concepts about the nature of the universe and of the gods in nearly every community in which the ethnographic research has been penetrating.
>
> The deification of important aspects of nature continues as a crucial feature of the religious symbolism: the sun, the moon, rain, and maize are all prominent in most contemporary Maya belief systems. There are also usually one or more types of underworld earth gods. The sun is often associated with "God" in areas where Catholic influence is very strong in the theology. Even more common is an association of the Virgin Mary with the moon goddess, as in Zinacantan. Rain is believed to be controlled by various types of essentially aboriginal gods—for example, Yahval Balamil (the Earth Lord) in Zinacantan.
>
> VOGT 1969:599–600

In the same region, a similar situation has been found regarding settlement patterns, social structure, kinship system, political organization, ancestral deities, and other reminiscences of pre-Colonial life. Although tinted with the color of imported ideas and practices, they still demonstrate their native origin and deep-rooted persistence (Villa Rojas 1947; Vogt 1969).

It must be noted that the vitality of these ancient cultural traits varies markedly from one region to another since the acculturation process has developed at diverse tempos in different places within the area. Generally, it is accepted that the lowlands, with greater communication facilities, have been the most acculturated and as a consequence live closer to the modern way of life. A contrast is offered by the highlands of Chiapas and Guatemala, where multiple mountain barriers have contributed toward maintaining the isolation and conservatism of Indian groups. In the highlands of Chiapas live in great numbers the Tzeltal and Tzotzil whereas, in the mountainous country of Guatemala, are enclaves of communities speaking Chuj, Ixil, Jacaltec, and Mam, all languages of the Maya family. It is among these groups that one still finds, with a highly pronounced vitality, the most conspicuous survivals of the ancient tradition.

Although undoubtedly these are the least affected areas of the entire Maya realm, they have not remained entirely free from cultural influences introduced by the whites. The diverse stages of this acculturative process, as it occurred in the Cuchumatan region of Guatemala, have been summarized by Oliver La Farge (1962:281–291) in the following way:

1524–1600 *Period of the Conquest:* This was characterized by violent changes which shattered the indigenous cultural structure.

1600–1720 *Period of the Colonial Indian:* The abolition of the *encomienda* and of forced labor, permitted the natives to lead a more tolerable mode of life. During this period Hispano-Christian elements were absorbed in a larger proportion and altered to some extent. Many Maya elements were destroyed, mutilated, or heavily changed.

1720–1800 *First transition period:* Gradual weakening of Spanish control, with the emergence of ancient Maya elements, which had been long repressed. Integration of Maya and Christian-Spanish elements into a new pattern more suitable to the Indian way of life.

1800–1880 *Recent Indian period I.* Characterized by a better adjustment with the new system, social equilibrium and a slow internal evolution, without strong external pressure.

1880–?? *Recent Indian period II.* A new wave of external intervention triggered by the incipient coffee industry which needed new arable lands. Influences of the machine age and of the Latin American culture are felt. The process of acculturation and social conflicts are accelerated.

What has been said about the Cuchumatan zone in Guatemala can be extended to the Chiapas highlands, which present marked cultural affinities with it. It must be repeated that, in spite of the historical happenings mentioned by La Farge, these two zones constitute the most conservative region in all the Maya area and exhibit the greatest degree of persistence in not a few practices, customs, and concepts dating from pre-Spanish antiquity. Our information will frequently proceed from this zone in our search for data related to the aboriginal concepts of time and space.

The earth crocodile and the concept of space

In Dr. León-Portilla's study it has been seen that one of the oldest Maya concepts concerning the spatial image of the earth's surface was that of a monster with the jaws and claws of a crocodile or with the form and head of a fantastic reptile. According to Thompson (1950:12), this idea of the saurians or reptiles representing deities of the earth or rain constitutes one of the most prominent motifs in the ancient Maya mythology and religion. This idea is similar to that of the *Cipactli,* a crocodile figure, that represented the earth's surface in Aztec cosmology.

Among the Yucatec Maya, the supreme god *Itzamná* was usually conceived of as having the form of an alligator or, also, as a serpent with two heads. In its terrestrial impersonation this deity was called *Itzam-Cab-Ain,* which Roys (1933:101) has translated as "Itzam-Earth crocodile." Memory of it has endured among the Maya until relatively recent times in which the *Books of the Chilam Balam* were still consulted. In the *Pérez Codex* (1949:231), which contains part of the *Chilam Balam of Maní,* there is a reference to the way in which the earth's surface was formed from the body of *Itzam-Cab-Ain,* the "Monstrous alligator." The text is as follows:

> . . . when there was a great cataclysm, when the monstrous alligator arose, this occurred at the conclusion of a series of *katuns.* With a deluge, time will come to an end. The 18-*Bak katun* was counted in its seventeenth part, before the *katun* terminated. *Bolonti Ku* did not wish the monster to destroy the world, for which he cut its throat and formed the surface of Peten.
>
> PÉREZ CODEX 1949:231

It is worth mentioning that, according to the *Motul Dictionary,* the term *peten* signifies "province, region, or district" as well as "island." Apart from texts such as this, and other fragments of cosmogonic content, still preserved by native believers in the old traditions, the names of *Itzamná* and of his various representations remained practically forgotten throughout all the Maya area. At present, *Itzamná* is not mentioned at all in the life and beliefs of the Yucatec Maya.

An exception to this is the case of the Lacandon Indians of the state of Chiapas, descendants of the Yucatec Maya, who for centuries remained isolated from all exterior contact. It was among these that Tozzer discovered (1907:96) the belief in a god called *Itzana* whose residence was in the old ceremonial center of Yaxchilán. The author himself points out that the similarity of the name of this god with that of *Itzamná* is very striking. He adds that in the Lacandon pantheon this god's position is much lower than that formerly held by *Itzamná*. No less interesting is the fact that some of his informants told him that *Itzana* was the Guardian of the Underworld.

On the other hand, Tozzer mentions another god called *Itzananchqu,* a name that seems unintelligible at first but once its components are analyzed, it becomes extremely significant. Its elements are *Itzan-Noh-Ku,* which is like saying "Alligator-Great-God." The value of what is suggested by this name is increased when we consider that in the highest part of the cliff where he has his abode (next to the lake of Petha) is carved a two-headed serpent which, according to Thompson (1960:117), at times represented *Itzamná.*

When the Indians were asked for an explanation of this figure, they simply answered ". . . that it was done by the god who inhabited the cliff" (Tozzer 1907:69). The cult of this "Alligator-Great-God" is still alive and constitutes one of the most serious religious obligations of the natives inhabiting the district where the sacred cliff is located. Tozzer had the opportunity to accompany one of the pilgrimages that visited the sanctuary although he was not permitted to witness the rites that took place in the interior of a deep cleft or cave which is at the foot of the cliff and is considered to be the entrance to the house of this god. Philip Baer, a North American missionary and linguist who has lived for decades among these Indians, assured us in a recent conversation (1967) that even today they continue to practice this cult. He added that at the rear of that cave have been found human bones which could have been those of notable ancestors of one of the clans, left there as was the custom of the ancient Maya. (Concerning the presence of ossuaries in this region, see Blom 1954 and Soustelle 1937:52–53.)

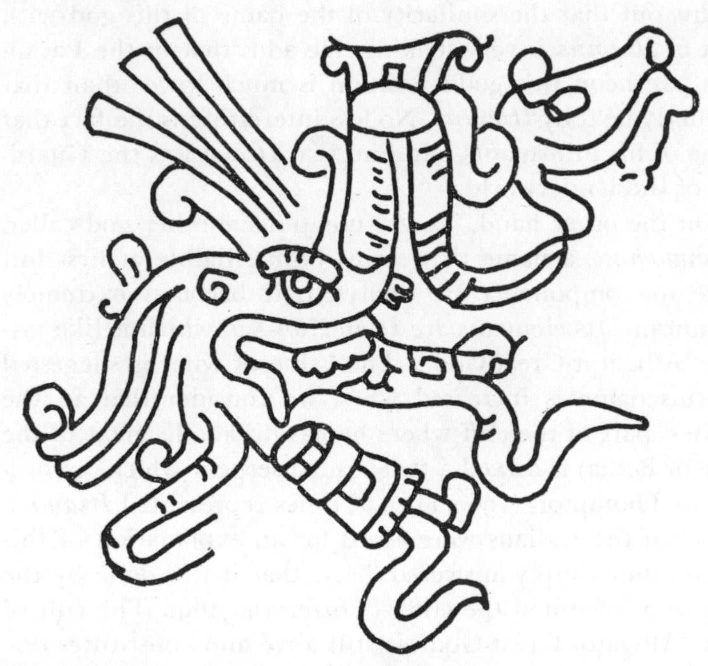

Appendix Figure 1. Two-headed serpent carved on the cliff where the god *Itzan-Noh-Ku* is worshiped.

To conclude with the theme of the *Itzam,* the supernatural being which represents the surface of the earth, certain suggestive details gathered by Thompson among the Maya of San Antonio, British Honduras, are worth mentioning. The natives venerate the *Mams* as deities of the underworld, mountains, valleys, thunder, lightning, rain, as well as of fishing, hunting, and agriculture. In many aspects the *Mams* closely resemble the *Chacs,* who also have survived in the beliefs of the Yucatec Maya. At times the *Mams* are considered to be a single being; at others, four; and at yet others, a legion, since there are *Mams* everywhere and with varying functions. There is no clear idea concerning their appearance; sometimes they are thought of as old men wearing sandals of moleskin and sitting on chairs made of armadillo shells.

There are four *Mams* of major importance who direct the others. These have special names and colors and are located at the four cardinal points. The one to the west, who is highest in rank, is white. The rest are colored yellow, red, and black. The *Mam* corresponding to the eastern zone is called *Itzam.* He is said to rule over the hot regions. In the prayers still directed to him, when sowing is begun, he is called *Santo Itzam,* and his devotee tells him, "I am going to wound your breast." Concerning the meaning of this phrase, Thompson points out that because the body of *Mam* is the earth, to make holes in it for planting seed may very well be described as beating or wounding the breast of *Mam* (Thompson 1930:51). There also exist versions in which *Mam* is bound in the underworld and that, on moving himself from time to time, produces those strange noises that may be heard in the bowels of the earth at the start of the rainy season; that is, when the sowing is begun. To sum up, although the idea of the alligator or crocodile has now disappeared from among the Maya of British Honduras, the term *Itzam* has been preserved in reference to a deity representing the earth's surface, such as appears in the previously reproduced passage from the *Pérez Codex*.

The fourfold image of the universe

Contrasting with the idea of the monstrous crocodile that forms the earth's surface and which has practically disappeared from the cosmogony of the modern Maya, there exists the concept of the fourfold image of the universe as a surviving element throughout the entire Maya area. The historical antecedents of this concept, studied through the stelae, panels, codices, and Colonial texts, have already been analyzed in the preceding pages by Dr. León-Portilla. We will now demonstrate the vitality of that concept in the thought and acts of the groups of natives we are dealing with. As we will see, as in ancient times, the contemporary Maya continue to have a fourfold vision of the world in which the terrestrial surface is distributed into four great sectors that part from an imaginary center and extend toward the four points of the compass. Each of these has its own gods, colors, fates, and, moreover, specific connections with the diverse cycles that form the aboriginal calendar. It should be noted that the vitality of these ideas is of uneven distribution, for in some areas certain aspects are preserved while in others they have been lost, a greater emphasis being placed on certain details. In the lowlands, for example, the notion of the native calendar has disappeared completely and therefore, the chronological significance of each of the diverse cosmic sectors has also been lost. On the other hand, many of the ideas integrating the fourfold model of the universe have been preserved with considerable strength. A tangible consequence may be found in the model coresponding to the layout of the village, of the cornfield, of the pagan altar, as well as in the ritual of diverse non-Christian ceremonies.

As can be recalled, the basic lines of this structural pattern were recorded in the *Popol Vuh:*

> Great were the descriptions and the account of how all the sky and earth were formed, how it was formed and divided into four parts; how it was partitioned, and how the sky was divided; and the measuring-cord was brought, and it was stretched in the sky and over the earth, on the four angles, on the four corners, as was told by the Creator and the Maker, the Mother and the Father of Life. . . .
>
> POPOL VUH 1961:80

Once these four great divisions of the world were established, their representatives or regents were placed in them and each was given his specific color: red to that of the east; white to that of the north; black to that of the west; yellow to that of the south. On speaking of these directional gods, Landa says that

> Among the multitudes of gods which this nation worshipped they worshipped four, each of them called Bacab. They said that they were four brothers whom God placed, when he created the world, at the four points of it, holding up the sky so that it should not fall. They also said of these Bacabs that they escaped when the world was destroyed by the deluge. They gave other names to each one of them and designated by them the part of the world where God had placed him, bearing up the heavens. . . .
>
> LANDA 1941:135–136

The other names with which they were designated were the following: the one to the east was called *Chac-Pahuatun* or *Chac-Xib-Chac;* the one to the north, *Zac-Pahuatun* or *Zac-Xib-Chac;* the one to the west, *Ek-Pahuatun* or *Ek-Xib-Chac;* and the one to the south was called *Kan-Pahuatun* or

Kan-Xib-Chac. The particles *Chac, Zac, Ek,* and *Kan,* with which each name begins, simply signify the colors red, white, black, and yellow.

In our attempt to see up to what point these ideas are still a part of the daily life of the contemporary Maya, what we found in the village of Chan Kom, Yucatán, during our long residence there, is worthy of mention. In the first place, the village itself is conceived as a quadrilateral having four principal entrances oriented toward the four cardinal points. At every one of these entrances one encounters two piles of stones facing each other, each with its own cross. Regarding this point, Landa tells us that

> It was the custom in all the towns in Yucatan that there should be two heaps of stone, facing each other at the entrance of the town, on all four sides of the town, that is to say, at the East, West, North and South. . . .
>
> LANDA 1941:139

These four entrances in modern times are thought of as the "four corners" of the villages. It is believed that four supernatural beings, called *Balams,* arrive at nightfall and set themselves at each of the corners. Their function is that of guarding the quadrilateral formed by the village, preventing the intromission of malevolent spirits that might alter the well-being of the inhabitants. In the center of the community is set a fifth *Balam,* known as *thup* (little one), who, in spite of his small size, is considered to be the most powerful. These five supernatural protectors are called *balamob caob* (guardians of the town). In the center of the distant communities, one can still see the "tree of abundance," represented by the always green and leafy ceiba. Actually, the village of Cham Kom has seven entrances, in spite of which the people only take into consideration the four principal ones that form the "corners" of the towns. The same idea is applied to the cornfield, always of quadrangular form, with its four corners and central point which are thought of as the seats of the five *Balam col* (guardians of the cornfield). These prevent the entrance of animals and adverse forces that could damage it (Redfield and Villa Rojas 1962:114).

This spatial idea of a quadrilateral with four points marking its "corners" or entrances is so rooted among the lowland Maya that even among the Lacandon, who do not have compactly organized towns, one finds the same concept (Tozzer 1907:39).

This image of the terrestrial plane is formed in the same manner as the ancient one although in it the gods of the wind *(Pahuatuns)* and of the rain (*Chacs*) occupy its cardinal points. Those versed in the ancient traditions consider them to be of the same nature and, therefore, in their prayers employ both names interchangeably. They also invoke the four great *balams* that are in the four corners of the heavens, in the four corners of the clouds.

In the ceremony of the *Chac-Chaac,* which takes place in the brush, asking rain of the *Chacs* and *Pahuatuns,* the four-sided rustic altar constructed of tree trunks represents the terrestrial plane. Distributed in its four corners are four individuals personifying the "great *Chacs,*" and, moreover, in order to attain the number five, another one is located toward the east, at about ten meters from the altar. The latter is called *Kumku-chac* (thunder-chac). He is said to reside in the *chun-caan* (the trunk of the heavens) and to be of the highest rank. All these personages carry as the emblem of their office a wooden machete with which they pretend to produce lightning; they also carry a calabash filled with water for sprinkling on the altar. As might be expected in this ceremony, as in other native ceremonies, they also invoke the Christian saints, the cross, and the *Hahal-dios,* the "catholic God" (see Redfield and Villa Rojas 1962: 138–143).

Concerning the highland Maya, it is worth noting that they have the same vision of the terrestrial square and of the cardinal points, although their daily life is less influenced by it. In speaking of the Tzotzil of Larrainzar in Chiapas, the anthropologist William R. Holland (1963:69) states that "for them, the earth is the center of the universe, being a square, level surface, upheld by a bearer in each corner" (Holland 1963:69). Further on, he adds that "The Tzotzil conceive of the world as a square; heaven and earth being united at the corners. The *Kuch Vinagel-Balumil* are the gods of the four corners that support the world on their shoulders. Their slightest movement will produce tremors and earthquakes" (Holland 1963:92). The same author tells us that these "bearers of the world" do not correspond to the four points of the compass but to the intermediate points. Another four supernatural beings are to be found at the cardinal points. Each of these performs a function related to life and each is associated with a color. The yellow god, in the east, sends rain; the white god, in the north, is the deity of maize; the red god, in the south, blows wind; and the black god, in the west, sends death. Taken together, the gods that support the world and those of the four cardinal points are designated *Vashak Men*. On the other hand, in Zinacantan, a community not far from Larrainzar, also inhabited by Tzotzil Indians, it is believed that these *Vashak Men* are only four and are situated in the cardinal points. Some consider them one of the manifestations of the god that created the world, he who gave life to all that exists on earth: the springs around which the group lives, the mountains and even the churches, Tzotzil ceremonial centers, where the creator installed the saints (Vogt 1966:91–92).

With respect to the indigenous groups inhabiting the zone of the Cuchumatan in Guatemala, their concept of the terrestrial plane follows the same lines: it has a quadrangular form and deified beings dwelling in the four cardinal points. In this Guatemalan area, the four cosmic sectors are intimately related to human destinies, according to the "burden" corresponding to the diverse dates of the aboriginal calendar. We also know that at Nebaj, an Ixil community in Guatemala, groups of four crosses are set up upon or near archaeological mounds. They are ordered according to the four corners of the heavens and are considered to be bearers of the year.

The corners of the world

The fact that in Larrainzar the Tzotzil Indians believe that the four corners of the world are situated in the intermediate points of the compass can lead to a brief digression. It concerns the possibility that the ancient Maya may have given cosmogonic meaning to those points, aside from the importance always accorded the cardinal points as appears in the stelae, codices, and Colonial Indian manuscripts. Actually we had discovered the same idea among the Maya informants in various parts of the lowlands. Thus in 1932, an aged *Ah-kin,* the last *Batab* or governor of Chemax, in eastern Yucatán, emphatically told us that the four corners of heaven corresponded approximately to sites located between the four intercardinal points. In accordance with this, it should be remembered that in the aforementioned ceremony of *Chac-chaac* the four personages representing the *Chacs,* rain gods, are situated in the four corners of the altar which point approximately toward the intermediate points of the compass.

On studying, at a later date, the indigenous group at X-cacal, in the southeast of Yucatán, we found that the sacred capital, in which a "talking cross" was venerated, was oriented in a highly significant manner. Generally the natives called that place *Santo Cah* (sacred town) instead of

X-cacal, its proper name. The principal temple there, built of palms, wattle, and daub, occupied the center of a large plaza. Next to it was a large edifice of the same material, used for public meetings, feasts, and large communal gatherings. In its function, it was very similar to the *Popol-na* (town hall) of the ancient Maya villages in which, according to the *Motul Dictionary,* the people gathered to discuss issues of common interest or to practice dances for the village fiestas. These two buildings were enclosed within an imaginary square of fifty meters per side, the corners of which were indicated by four crosses on top of small mounds of stone. The space thus delimited was protected in this way from evil winds and other adverse elements that roam the world. These four crosses corresponded to the intermediate points of the compass. Upon inquiry about the meaning of that spatial distribution, the native priests informed us that the church and the crosses around it were ordered according to the model established by God when he set up the first holy place here on earth. The central point (now occupied by the temple) was called *Xunan-Cah* (Virgin-village) because the altar of the virgin had been there. The points where the crosses are presently located received the respective names of *Belem-Cah* (Bethlehem-village), *Cah-Paraiso* (Paradise-village), *Cah-Jerusalem* (Jerusalem-village) and *Xocen-Cah* (Saint Joseph-village). The ancient cosmic pattern of five points is perceptible here though mingled with Catholic ideas and beliefs (see Villa Rojas 1945:43).

In respect to the origin of these five directional points, this could be attributed to those indicated by the sun, in its rising as well as its setting, in the moment of its maximum declination at the solstices. One would thus have two points corresponding to the rising and setting of the sun on June 21 (summer solstice), and two more resulting from the rising and setting of the sun on December 21 (winter solstice).

The fifth point would represent the moment when the sun passes the zenith. Expressing this in a pictorial form, one obtains the following figure:

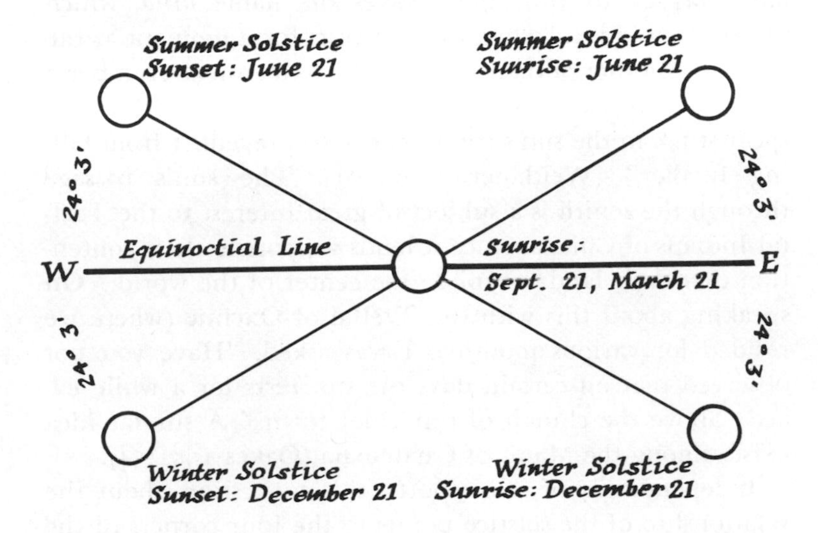

Appendix Figure 2. Position of the four "corners of the heavens" as related to the sun in its solstices, and of the central point as the sun passes through the zenith.

It should be remembered that those four points situated to the sides of the equinox line are no further than 24°3′ from it, either to the north or to the south. That is to say they lack more than twenty degrees to arrive at the midpoint between the extremes of the compass. They do not, therefore, form a square but, rather, a rectangle with two corners toward the east and the other two toward the west.

These solstice points were of basic importance in the observations of Maya astronomers, to the point that they built complex observatories for the purpose of determining the exact moment of the solstice throughout the course of the year. Even today there are traces, among some indigenous groups, of a preoccupation with the extreme points marked by the sun at seasonal changes. Among the Chinantec of Oaxaca, for instance, the sixth month of their calendar (May 21 to June 9), receives the name *hinà,* which means, at midday, "when the sun rests for a moment at the zenith." Another of their months, the sixteenth (December 7 to the 26), is called *hi ta nyi,* a term signifying "thorn or spear stuck in the sun's countenance to prevent it from falling further" (Weitlaner 1936:201). The sun's passage through the zenith is a subject of great interest to the Tzeltal Indians of Chiapas since it lends support to their contention that their land occupies "the center of the world." On speaking about this with the Tzeltal of Oxchuc (where we resided for various months), I was asked, "Have you not observed that on certain days the sun rests for a while exactly above the church of our chief town?" A similar idea exists among the Mams of Guatemala (Oakes 1951:54).

Independently of our reports, this suggestion about the relationship of the solstice points to the four corners of the heavens has been discussed on other occasions. For example, Vogt mentioned it in a meeting of Maya specialists celebrated in 1962, though in a context that nullified the meaning of the cardinal points. His own exposition was the following:

> The Maya spatial orientation of the four corners of their universe is not based upon our cardinal directions of north, south, east and west, but probably to either our intercardinal points (northeast, northwest, southwest and southeast), or toward two directions in the east and two in the west, that is to say, sunrise at winter and summer solstices, and sunset at the same two solstices.
>
> VOGT 1964:390

As may be supposed the terms of Vogt's proposition gave rise to disagreement at the above mentioned meeting due mainly to the little importance which he gave to the cardinal points which, in our opinion, had very deep meaning in the cosmogonic concepts of the ancient Maya.

Another version of the same idea was presented by Rafael Girard, based on data obtained from a native Chorti priest in the town of Cayur, Guatemala. Essentially he argues that:

> From the point of view of the Chorti the four corners of the cosmic quadrangle correspond to the extreme points of the visible horizon where the sun rests in its regular annual oscillation. That is to say, in modern astronomic terms, they correspond to the points of the ecliptic at which the sun arrives in its maximal and minimal declinations in the celestial parallels corresponding to the tropics.
>
> GIRARD 1962:45

The same author affirms that even today the Chorti continue to record the movements of the sun in order to establish the points of the solstice with precision. He adds that each angle of the world where the sun pauses is indicated at its corresponding corner by a stone marker, thus showing the limits of the cosmic plane.

The Maya of Yucatán left evidence of their ancient interest in these points, in one of the passages of the *Chilam Balam of Chumayel,* on dealing with dates which corresponded to said points in the Julian calendar. In his characteristic style, the Indian scribe asserts that:

> When the eleventh day of June shall come, it will be the longest day. When the thirteenth day of September comes, this day and night are precisely the same in length. When the twelfth day of December shall come the day is short, but the night is long. When the tenth day of March comes, the day and night will be equal in length.
>
> ROYS, *Chilam Balam of Chumayel,* 1933:86–87

The fact that the dates cited are anterior to the Gregorian correction of 1582 reveals the antiquity of this annotation. Turning to the pre-Columbian sources we find in the *Dresden Codex* figures that, according to Seler (cited by Spinden 1957:91), represent the sun. Said figures correspond to the idea we are trying to establish: that the Maya took into consideration both the cardinal points and the intermediate points. We reproduce one of the *Dresden* illustrations here (Figure 3).

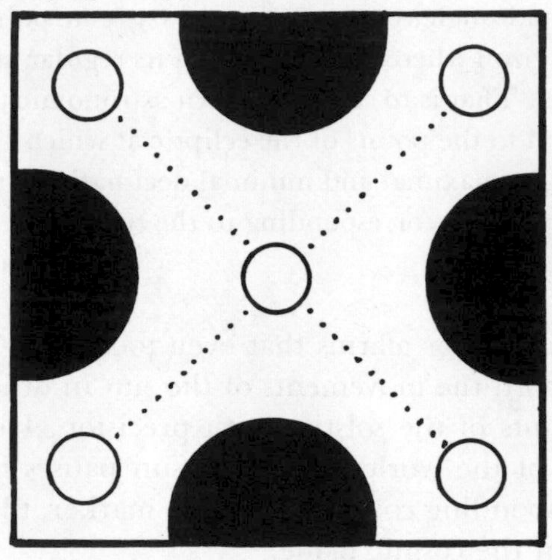

Appendix Figure 3. Symbol of the sun according to the *Dresden Codex.*

Of even greater significance is a figure in the *Fejervary-Mayer Codex,* from the highlands of central Mexico, in which the four sectors of the world appear precisely distributed toward the four directions of the compass and surrounding the fifth zone occupying the center. At the intercardinal points that form the corners are found symbols of calendrical significance. The figure is rich in cosmic symbolism and, according to Roys (1933:100), corresponds largely to what the *Chilam Balam of Chumayel* states about the creation of the world (Figure 4).

Although other examples could be adduced to confirm the existence of this interest of the Maya in taking into consideration the cardinal as well as the intermediate points, the above digression seems sufficient.

Another Maya spatial concept, now beginning to emerge

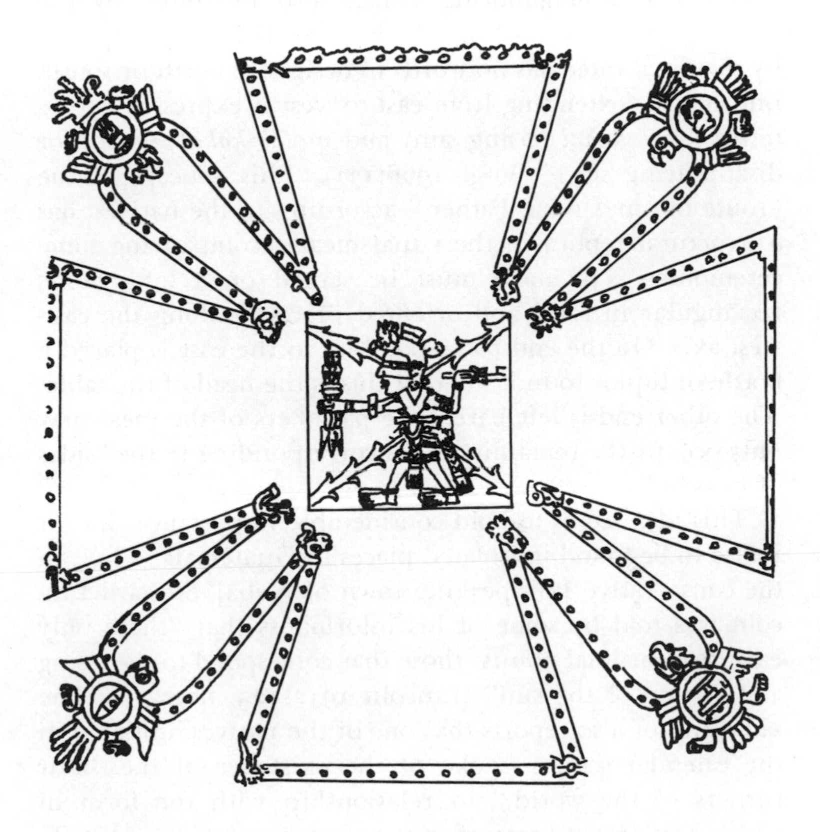

Appendix Figure 4. Diagram of the four cardinal regions and of the four corresponding corners. The entire figure is dedicated to *Xiuhtecuhtli,* Lord of Fire (*Fejervary-Mayer Codex,* 1). According to Roys, the figure corresponds to the concept of the cosmic quadrangle described in the *Chilam Balam of Chumayel* (1933:100). See also pages 75–76 of the *Madrid Codex* reproduced in Figure 19 of the first part of this work.

from ethnographic studies, recognizes only two cardinal points, east and west, while the north and south are merely said to be "the sides of heaven." This information was gathered at the Tzotzil community of Chenalo, Chiapas, by Calixta Guiteras (1965:38). Something similar was found in Zinacantan, a neighboring village also inhabited by the Tzotzil. According to Vogt: "The Tzotzil language spoken by the Zinacantec has no words to designate north or south, but the axis extending from east to west is expressed by the terms *lokeb-kakal* (rising sun) and *meleb-kakal* (setting or disappearing sun)" (Vogt 1966:131). This concept of the "route of our Divine Father," according to the natives, has a structural replica in the ritual meal accompanying some ceremonies. The meal must be served on a long table, rectangular in form and oriented precisely along the east-west axis. On the end corresponding to the east is placed a bottle of liquor to indicate that this is the head of the table. The other end is left bare. The partakers of the meal may only occupy the remaining sides corresponding to the "sides of heaven."

This idea seems to hold considerable importance since it is also to be found in isolated places in Guatemala. Thus, in the conservative Ixil-speaking town of Nebaj, Steward Lincoln was told by some of his informants that "there only exist two cardinal points: those that correspond to the rising and setting of the sun" (Lincoln 1942:125, note 13). The same author also reports that one of the natives, initiated in the calendar secrets, spoke of the existence of the "four corners of the world," in relationship with the form in which the days representing the "yearbearers" are distributed in space (Lincoln 1942:110).

As mentioned above, these aspects of the surviving Maya groups concerning horizontal space are only beginning to be known. A greater amount of information will be required in order to grasp all its implications.

The vertical dimension of the universe

Up to this point our attention has been focused exclusively on surviving ideas concerning the horizontal image of the universe. Now it is necessary to consider the ideas corresponding to its vertical structure with its thirteen heavens and nine lower levels, so prevalent in Central America. Archaeologists have pointed out the antiquity of these concepts, since the glyphs of the nine gods of the underworld as well as certain allusions to the deities of the upper levels appear in Classic inscriptions. Moreover, as shown by León-Portilla in the present work, in the ancient city of Yaxchilán were found vestiges of nine figures which experts consider to represent the nine gods of the underworld; a similar discovery was made in one of the temples of Palenque.

This vertical dimension, with its two divisions of nine and thirteen floors, continued to thrive in the interpretation of cosmic phenomena far into the eighteenth century, when the major portion of the texts comprising the *Chilam Balam* books was composed. In the most important of these, that of Chumayel, are mentioned repeatedly the conflicts existing between *Oxlahuntikú* (Thirteen gods) and *Bolomtikú* (Nine gods). These struggles were so violent as to cause floods and even the fall of the heavens. In the same text these nine gods of the lower world are conceived of as the authors of evil and sin when these first appeared on earth (Roys 1933:105).

In respect to the disposition of the thirteen heavens, they could have been arranged in a strictly vertical manner, one above the other. Or also, as Thompson suggests, in the form of a double ladder with six steps ascending the eastern side and six descending the western side, the thirteenth corresponding to the zenith which unites the two sides (Thompson 1954:225). Concerning this point, Krickeberg has given a precise description of the cosmic structure as follows:

> The cosmic image of the ancient Nahua tribes of Texcoco, Chalco, and Tlaxcala, also prevalent among the Maya, distinguished nine underworlds and as many heavens, while the Aztec sometimes spoke of thirteen heavens. According to the original concept, these heavens were the steps by which the sun ascended from east to west during the morning, and by which he descended in the afternoon in order to effect a parallel journey at night when passing through the kingdom of the dead. In this way, the highest heaven and the lowest underworld were not to be found at the end but in the center of the two series of steps. The thirteen gods of the diurnal hours and the nine gods of the nocturnal ones correspond to the steps or levels of the heavens and of the underworld. The sun god reigns over the central (seventh) hour of the day, and the god of death rules over the central (fifth) hour of the night. Once again, spatial and temporal concepts have been coordinated.
>
> KRICKEBERG 1961:131

These two cosmological visions, the one of vertical order and the other in the form of a double ladder, have perdured to the present, and contemporary ethnographers have encountered them among various groups of the Maya area, in the lowlands as well as in the highlands. Tozzer, for example, cites the version he found among the Yucatec Maya: there, the celestial planes are seven, vertically disposed one above the other and each with a central hole through which passes the leafy *yaxché* (ceiba tree), that extends its branches over each floor. The souls of the dead go up these branches, according to their merits, in order to arrive at the highest plane where the Christian god resides. Other informants indicated that the ascension takes place by means of a

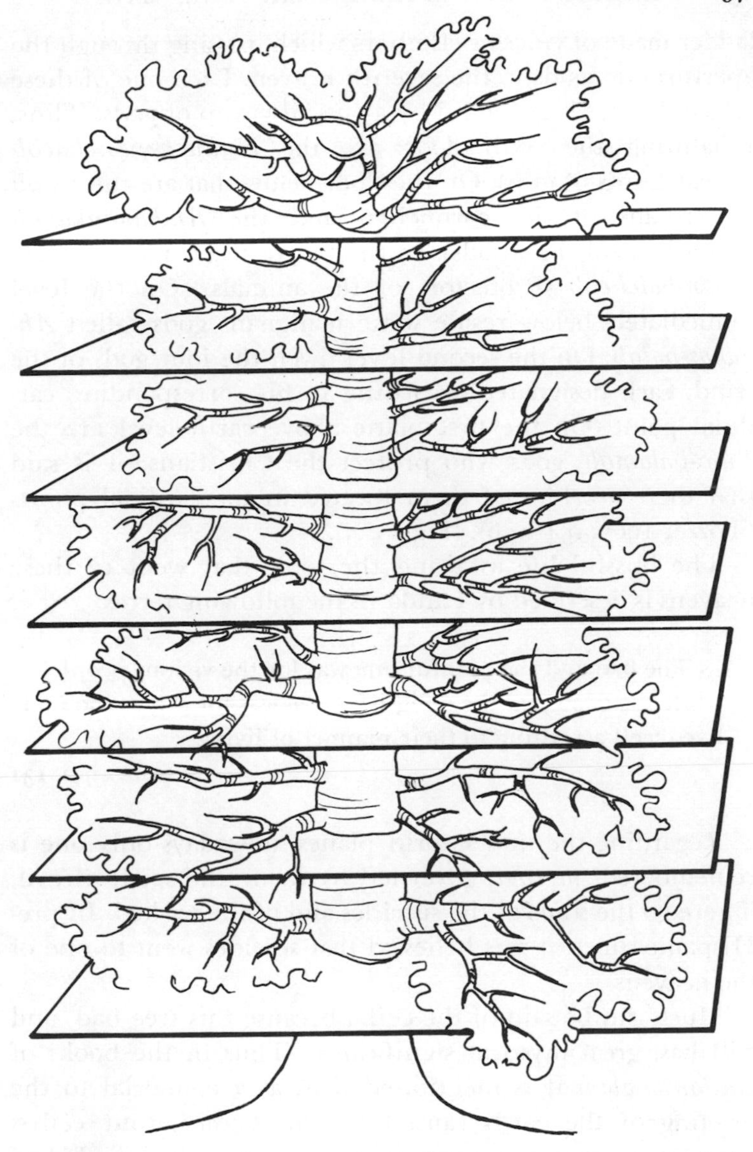

Appendix Figure 5. Image of the celestial levels and of the sacred ceiba tree, according to data obtained by Tozzer among the Maya of Yucatán.

ladder made of vines or climbers which, passing through the apertures, extends to the seventh heaven. Each one of these celestial floors is governed by a special group of gods. Thus, inhabiting the sixth floor are the *Nukuch-yumchacob* (Great Lords of rain). On the floor below that are the *Kuob* (Guardians of the cornfields) and the *Ah-kanankaxob* (Guardians of the woods). In the fourth heaven are the *Ah-canan-balcheob* (Protectors of the animals). On the level immediately below reside some malignant gods called *Ah-kakaz-balob*. On the second level dwell the four gods of the wind, each designated according to his corresponding cardinal point. On the first plane above earth level are the *Yum-balamob,* gods who protect the Christians; it is said that there are four of them located in the cardinal points (Tozzer 1907: 154–156). (Figure 5).

The blissful life awaiting the souls that went to these heavens is described by Landa in the following terms:

> The bad and the painful one was for the vicious people, while the good and the delightful one was for those who had lived well according to their manner of living.
>
> LANDA 1941: 131

Regarding the underworld planes, nowadays only one is remembered: *metnal,* governed by *Kisin,* the spirit of evil. There go the souls of the suicides and of the wicked. In pre-Hispanic times, it was believed that suicides went to one of the heavens.

Much can be said of the ceiba because this tree had, and still has, great mystical significance. Thus in the books of *Chilam Balam* it is mentioned that, as a memorial to the leveling of the earth caused by the *Bacabs,* four ceibas were set up at the four cardinal points and, as a symbol of the future age, a fifth one was placed in the central point. In the corresponding prophecy one reads:

> Also the *Yax Imixche,* green ceiba, will be raised up in the center of the province as a sign and memorial of the annihilation. It is the one that supports the plate and the vessel; the mat and the throne of the *katuns* live because of it.
>
> BARRERA VÁSQUEZ 1948:155

Among the Itza of Tayasal the central tree was adored as *yaax-chel-cab,* which means "the first tree of the world"; it was also called "the tree of life." Landa refers to it as *Vahomché,* the upright tree endowed with great power against the demons. The Franciscan fathers applied the same term to the Cross. Even today, according to Franz Blom,

> . . . the cross we see in their dwellings (or huts) is not the Christian cross but the direct descendent of this *Yaaxché,* the tree that demands water in order to flourish.
>
> BLOM 1956:284

Due to these surviving beliefs, one still finds an old leafy and pleasantly shady ceiba growing in the square of some Maya villages. In reference to the meaning of such ceibas, Núñez de la Vega, writing at the end of the seventeenth century, states:

> The ceiba is a tree which grows in all the village plazas within sight of the town hall. And under it they elect their mayors. And they offer it incense in braziers, and they believe firmly that their lineage came out of the roots of that ceiba and thus they have painted it on a most ancient canvas.
>
> NÚÑEZ DE LA VEGA 1702:9

Turning now to the highland Maya groups, one finds that the image of the universe among the Tzotzil of Larrainzar, Chiapas, is also stepped and pyramidal, similar to that described by Krickeberg as cited above. The version presented here was obtained from Indians themselves by the anthropologist William R. Holland:

> In the mind of the Tzotzil the earth is the center of the universe, being a square, level surface, held up by a bearer at each corner. This people envision the sky as a mountain with thirteen steps, six on the eastern side, six on the western, with a thirteenth in the middle forming the summit of the heavens. From below, the heavens resemble a dome or an upside down vessel placed on the surface of the earth. A gigantic ceiba rises from the center of the world toward the heavens. Imagined from outside of the earth, the heavens are an enormous pyramid or mountain. This concept of the mountain of the heavens is symbolized many times in the ancient codices. Under the earth is located the lower world, *Olontik,* the abode of the dead, which is composed of nine, thirteen, or an indeterminate number of steps.
>
> The Tzotzil, like the ancient Maya, consider the heavens the home of benevolent deities, creators and makers of all human, animal, and vegetable life. Conversely, the lower world is the residence of evil gods who eternally fight to undo the work of the heavenly gods and try to win over new occupants for the world of the dead. Life is a constant struggle between the forces of good and evil.
>
> HOLLAND 1963:69

According to the same Tzotzil group, the sun ascends in a cart, following a path of flowers, advancing one step per hour until it reaches the zenith, or heart of heaven, where he rests for one hour. Afterward, it renews its journey toward the western world where it disappears in the sea, leaving everything in darkness. Its passage through the underworld follows an inverse course: its first stage is a descent toward a central point, midnight, the hour in

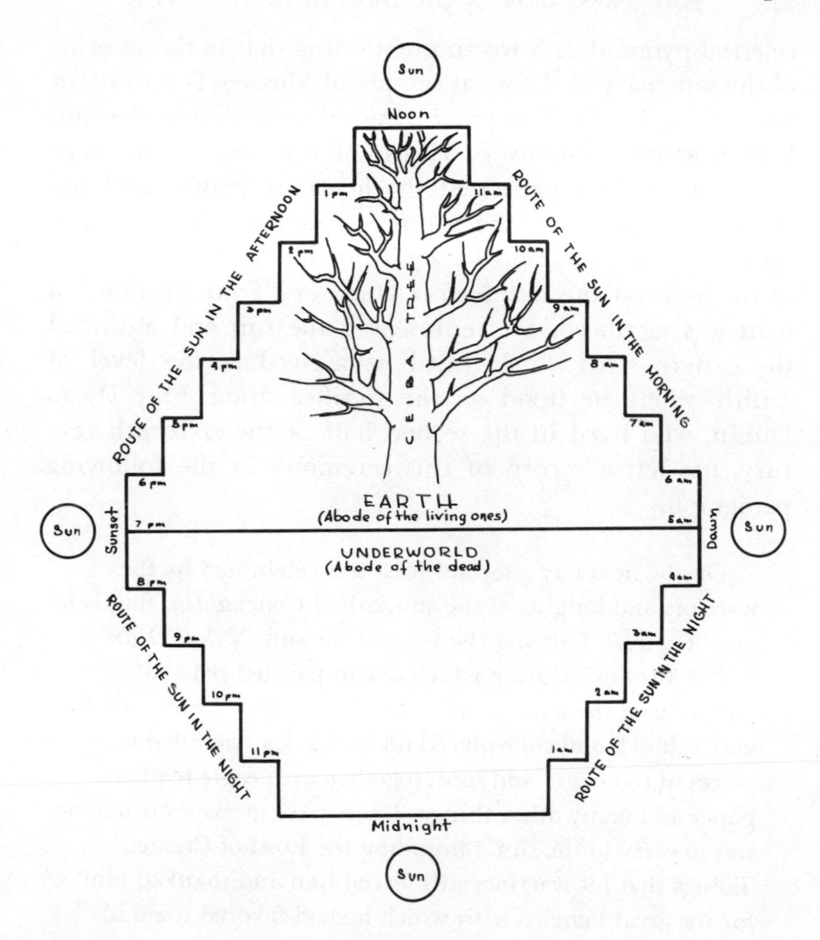

Appendix Figure 6. Spatial image of the universe, preserved to this day among the Tzotzil Indians in the state of Chiapas.

which it begins to ascend in order to arrive on time at its rising point and repeat its eternal pilgrimage (Figure 6). It must be added that, although good and evil gods exist, their distribution on the various floors of the heavens or of the underworld is not specified.

Confirming these beliefs about the image of a stepped

celestial pyramid, it is worth mentioning that in the interior of the sanctuary of the great temple of Mexico-Tenochtitlan there existed a small, stepped pyramid dedicated to the sun, known as the *Cuauhcalli,* House of the Eagle. The steps of this pyramid represented the heavenly planes, and the highest level—where the stone of sacrifice stood—corresponded to the hour of the zenith or midday. On the feast of the sun, on the day *Nahui Ollin* or "Four Motion," a man was sacrificed. He represented the sun and ascended the eastern steps slowly until he arrived at the level of zenith, where he stood on the sacrifice stone. Fray Diego Durán, who lived in the second half of the sixteenth century, has left a record of this ceremony in the following paragraph:

> On the next day a second feast was celebrated by the warriors and knights of the sun, called *Cuacuahtin,* that is to say, "Eagles." This was the feast of the sun, *Nahui Ollin,* "Four Motion," during which a man painted red was sacrificed in the name of the sun. They handed him a staff and a shield, and they placed on his back a bag filled with pieces of red ochre and soot, together with eagle feathers, paper and many other things. These were messages which he was to carry to the sun, reminding the Lord of Created Things that his warriors still served him and thanked him for the great benefits with which he had favored them in their wars.
>
> The victim, carrying the bag of gifts to the sun together with the staff and shield, slowly began to climb the steps of the pyramid. In this ascent he represented the course of the sun from east to west. As soon as he reached the summit and stood in the center of the great Sun Stone, which represented noon, the sacrificers approached the captive and opened his chest. Once the heart had been wrenched out, it was offered to the sun and blood sprinkled toward the solar deity.

Imitating the descent of the sun in the west, the corpse was toppled down the steps of the pyramid.

DURÁN 1964:121–122

According to Krickeberg, the ancient Maya represented those "solar steps" by means of miniature pyramids made of earth or wood and placed them in their temples on special festive occasions. As an illustration of how greatly diffused this practice is in Middle America the above author includes in his work a drawing of a wooden pyramid still to be found among the modern Huichol of Western Mexico (Krickeberg 1961:107).

The millenary persistence of methods for measuring the passage of time

On surveying the contemporary Maya ambit in search of extant traces of the ancient ways of measuring time, it can be seen that their persistence varies to different degrees. Thus, in the lowlands, in Yucatán, those ancient methods have almost disappeared. In the highlands of Chiapas, inhabited by the Tzeltal and the Tzotzil, a fair number of practices and beliefs connected with the calendar have been preserved. This survival is even more marked in the mountainous zone of the Cuchumatan, in Guatemala, where the *tzolkin* or 260 day-count, and the idea of the yearbearers continue to govern the destinies of the people.

In the case of the Maya of Chan Kom, in Yucatán, as in that of the inhabitants of Tusik, in Quinta Roo, the only survival is the ancient custom of counting the day starting at noon and not at midnight as we do. In Chan Kom, for example, we observed that the natives began their work in the cornfields beginning at midday on Sunday, since they considered that it was already Monday (Redfield and Villa

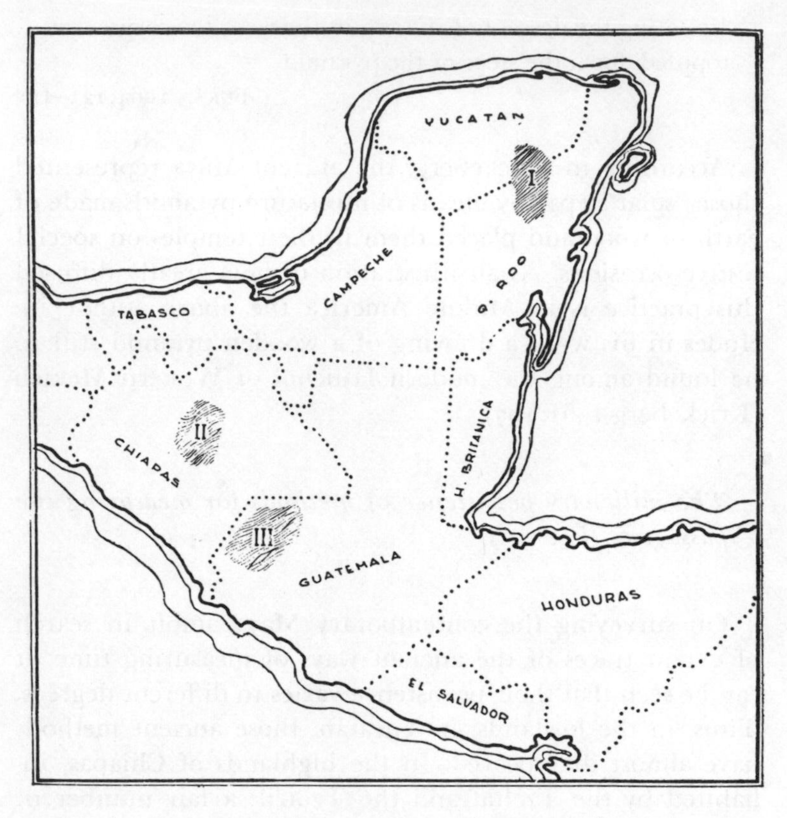

Appendix Figure 7. The least acculturated zones of the Maya area.
Zone I includes the villages of Chan Kom and Tusik.
Zone II shows the region inhabited by the Tzeltal and Tzotzil, and
Zone III is the area of the Cuchumatan, where the ancient Maya calendar is preserved to a certain degree

Rojas 1962: 184). We recorded a similar case at Tusik. The Indians at that place came to consult a Catholic almanac that we carried with us in order to know the saint's day upon which the child had been born so as to give him a name. The parents insisted that the name of the baby born after the hour of noon should be that corresponding to the next day (Villa Rojas 1945: 143–144).

This manner of counting the days was the same as that used by the ancient Maya. The Yucatec Maya scholar Pío Pérez reports that this people "placed the beginning of the year on the day the sun passes through the zenith on this peninsula on its way to the austral regions" (Pérez 1937: 534). It is probable that the Aztecs had the same system. In the opinion of Alfonso Caso "it seems highly probable that the ancient Mexicans did not compute the day from midnight to midnight, as we do, but from midday to midday, causing one Aztec day to correspond to two days of the Christian calendar and one Christian day to correspond to two Aztec days . . ." (Caso 1953: 103). Two sources confirm the idea. The *Telleriano Remensis Codex,* fol. 48 v., states that "they also count the day from noon of one day until noon on the next." The second source, Fray Juan de Córdoba, in reference to the Zapotec Indians of Oaxaca, notes that "they used to count the day from one noon to the next." (Córdoba 1942: 172).

In present times, the Mixe of Oaxaca follow the same pattern (Villa Rojas 1956:51). According to Steward Lincoln, some of the Ixil of Nebaj, in Guatemala, count the day from sunset to sunset and others from midnight to midnight. The same author adds that there is some confusion on this important point though in all cases Ixil ceremonies begin at sunset. (Lincoln 1942: 113).

In the highlands of Chiapas the ancient Maya solar calendar called *haab* (eighteen twenty-day periods plus the five extra days) is still preserved. Those five extra days were denominated "nameless days" because they passed without being taken into account. During our long stay among the Tzeltal Indians of Oxchuc (1943–1944), this form of measuring time was the only one in use. Any man or woman could tell me the date on this calendar but not its corresponding Christian day. The making of this correlation was the business of the native religious leaders, who were responsible for, the Church fiestas as well as for other sacred tasks. As in ancient times, the five extra days, also called *chay kin,* "the fiesta is lost" (Becerra, 1933:338), were considered unfortunate, and for this reason people tried to stay home in order to avoid accidents or encounters with underworld beings.

Apart from the religious significance of this calendar, its fundamental function at Oxchuc was of an agricultural nature, for by means of it the Tzeltal regulated all the activities in their cornfields. We will now record the names of the twenty-day periods, their equivalents in our calendar, and the tasks corresponding to each period or month, according to the information given us by the natives of Yochib at Oxchuc.

MONTHS OF THE TZELTAL CALENDAR

BATZUL — *26 December–14 January*
"Authorities are changed in the villages. Chili is sown, and a little brush is cleared or cut away; it is not yet a time for much work."

ZAKILAB — *15 January–3 February*
"The same as the previous month."

AGILCHAC — *4–23 February*
"The Carnival fiesta. Now commences the time for more work in clearing the brush."

MAC — *24 February–15 March*
"Clearing the brush. Sowing is begun in the cold upland."

OLATI — *16 March–4 April*
"Clearing. The moment to sow."

HUL-OL — *5–24 April*
"The same as the previous month."

CHAYKIN *25–29 April (five unlucky days)*
"No work is done. The *chaykin* period ends."

HOKEN-AHAU *30 April–19 May*
"Weeding is begun in the cold upland. Around here there still is not much work; some people spend their time scraping a little agave-fiber for making bag-nets and cord to be sold in Tenejapa."

CHIN-UCH *20 May–8 June*
"Weeding is begun in the cornfields around here."

MUC-UCH *9–28 June*
"Weeding and the performances of other minor tasks."

HUC-UINKIL *29 June–18 July*
"A little work in the cornfields clearing them of weeds. Houses are constructed. In the cold upland some brush is cleared."

UAC-UINKIL *19 July–7 August*
"Weeding is continued in the cornfields, for around here weeding occurs twice. The 6th of *Uac-Uinkil* is the exact date of St. James of Tenejapa." (Afterward, *ladinos* of Tenejapa told us that the 25th of July is the date corresponding to his feast.)

HO-UINKIL *8–27 August*
"Around here one is somewhat idle. In the cold upland they begin to clear the brush for next year's cornfield."

CHAN-UINKIL *28 August–16 September*
"Because it is not yet time to work here,

some people go to earn daily wages of 25 *centavos* in the cornfields of the cold uplands or in the coffee plantations of the Soconusco region."

OX-UINKIL *17 September–6 October*

"Just like the previous month. We have passed one hundred days of pure *Uinkils,* or idleness."

POM *7–26 October*

"Many busy themselves making ropes and bag-nets in order to sell them and then buy the candles and other things needed for the Day of the Dead."

YAXKIN *27 October–15 November*

"Time for building houses and chicken coops, for making bag-nets and other things. The 1st of *Yaxkin* is the Day of the Dead." (The natives hold these celebrations during the last five days of October whereas the *ladinos* celebrate them on the last day of October and the first two days of November.)

MUX *16 November–5 December*

"A little work is begun in the cornfields."

TZUN *6–25 December*

"Perhaps the 10th day of *Tzun* falls on the day of St. Thomas of Oxchuc." (This is the only date that does not fit in our correlation, for in the Gregorian calendar, St. Thomas' day falls on the 21st of December, which corresponds to the 16th of *Tzun*.)

With these data, taken directly from our field notes, the reader can form a fairly good idea of the manner in which the passing of the twenty-day periods regulates the Indian's work. Sánchez de Aguilar, grandson of Conquerors and one of the first chroniclers of Yucatán, in referring to the same calendar states:

> This count of the eighteen months plus the six canicular days [*sic* instead of five] is the same as our solar year of 365 days. It was highly useful to them, particularly for knowing the times in which to cut down the brush and burn it, and to await the rains, and to sow their wheat, maize and other legumes they sow at sundry times. And as the tillers of Spain observe such and such days and say, "Come October, make bread and cover," and other adages, thus, neither more nor less, was the usage and is the usage of these Indians in their sayings concerning these eighteen months and six canicular days for sowing and looking after their health and curing themselves, as we do in the Vernal, Estival, Autumnal, and Hibernal seasons.
>
> SÁNCHEZ DE AGUILAR 1937:142–143

Returning to the case of contemporary Chiapas, it must be explained that the informant's statement, "these are one hundred days of pure *Uinkils*," means that there are five periods of twenty days, the term *uinkil* signifying either *man* or period of twenty unities. The count of the *uinkils* is begun on June 9, at the entrance of the period *huc-uinkil,* which means "seventh period of twenty days." It is followed by the sixth, then the fifth, and continues thusly to the third. With respect to the meaning of the names of the twenty-day periods, a detailed analysis would be too lengthy, for which reason the reader is referred to the explanations given by Thompson (1960:109–119) and by Becerra (1933:61–70).

Apart from the time-count we have described, we found no pagan ceremony of a calendrical nature. The only act connected with the names of the twenty-day periods we recorded was an offering made in the period of *Chin-uch,* possibly a faint reminiscence of a more complex ceremony in honor of the god presiding over this period begging him to care for the cornfields. According to Professor Emeterio Pineda, this is the month in which the plants become infected, particularly by an insect, the aphid, that weakens and destroys them. At the present, this offering is simple and held in private. Each family hangs, at the entrance of its hut, a small bundle of the following articles in reduced size: a ball of *pozol,* dough, wrapped in a banana leaf, a red chile (*saja-ich*), a tortilla, an ear of maize, and a small stick similar to that used to remove ears of corn from the fire when toasted. This offering must be made on any one of the five days preceding the beginning of the month or, at latest, on the first day of *Chin-uch.* If it is made later, it is useless since it reveals that the farmer did not know how to comply adequately with his religious obligations. No prayer is said; it is sufficient to hang up the bundle. In connection with this rite, the natives anxiously await the passing of this month, for, if on the last day a sort of smoke appears in the atmosphere, it is believed that "*Chin-uch* burned its house" and that, as a result, there will be sufficient sun for good growth of the cornfields. If the smoke does not appear, the future becomes ominous because there will be too much moisture and the crop may fail. Although we tried to clarify the meaning of the phrase "*Chin-uch* burned its house," the natives could not give us more details. (Concerning the usages and customs of these Indians, see Villa Rojas 1946.)

In spite of the limited survival of the ancient Maya chronological concepts, the natives of the Chiapas highlands continue to be strongly attached to their manifest preoccupation over the ties that bind man to the eternal flow of time. In connection with this point, Evon Vogt, who has dedicated a number of years to the study of the Tzotzil at

Zinacantan, in Chiapas, has described the deep impression this cultural trait made on him.

> What I should like to add here is the extent to which our contemporary ethnographic data demonstrate that this preoccupation with time persists, even in places which no longer follow the ancient calendar. To cite some evidence from Zinacantan, which no longer uses the old calendar, each field season I am more impressed by the punctual manner in which our informants keep appointments made weeks in advance. . . . Again, I am impressed by the way in which the Zinacantec keep their successive turns in order on a waiting list for priestly duties for which they applied sometimes as many as twenty years in advance. Some individuals now know they will enter a certain religious office in 1982!
>
> VOGT 1964:35

Thus, although the forms of expression have changed, the ancient attitude toward the temporal dimension of human existence continues to operate as a determinant factor at all levels of daily life.

The computation of time, the gods and the rites

Turning our attention now to the south, in the direction of Guatemala, let us examine the third zone mentioned previously: that of the Cuchumatan. This zone occupies an ample section of the western part of said nation and is noted for its marked conservatism. Its cultural environment, predominantly Indian, gives the impression that time halted there many generations ago. The ancient calendar forms are so greatly preserved there that it has come to be called the land inhabited by the yearbearers' men (La Farge

Appendix Figure 8. Western Guatemala, a region in which the ancient Maya calendar has been preserved to some extent.

and Byers 1931:111). Located in this region are the communities of Jacaltenango, Chimaltenango, San Mateo Ixtatan, Todos Santos, Santa Eulalia, and others for which there exist extensive monographs treating this theme with a good measure of detail. The titles of these works may be consulted in the bibliography at the end of the present work. However, their conservatism notwithstanding, none of those places portrays with greater relief and clarity the existence of the ancient Maya practices and concepts (discussed above by León-Portilla), as do the three Ixil municipalities—Nebaj, Cotzal, and Chapul—located a few hours' walk to the east of the other villages. The region is still quite isolated, permitting usages and customs to exist with fewer alterations than in other places. From this region comes the best study yet realized on this fascinating theme of the survival of the ancient Maya calendar with its wealth of rites, beliefs, ceremonies, and divinatory practices which exercise such a profound influence on the life and fate of the natives. J. Steward Lincoln is the author of this remarkable work, *The Maya Calendar of the Ixil of Guatemala.* The premature death of the author, which occurred in the area he was studying, prevented him from writing the final draft. This task was realized by J. Eric Thompson, the foremost authority in this field. To gain an idea of the rich content of the Ixil calendar, let us cite the words of its discoverer, J. Steward Lincoln:

> This chronological register, which preserves many old Maya and old Quiché features not found in other current calendars, is not only a measure of annual solar time, with the exception of leap-year calculations, but a religious, ceremonial, and divine director of man's destiny. It controls his daily life in the spheres of worship, agriculture, domestic and social relationship; and influences his behavior in connection with birth, love, marriage, death and livelihood. The days, which are also divinities to whom he prays, exert favorable and unfavorable influences on all his activities. Although some of the features of this ancient calendar are almost lost, it still endures as the core of Indian religion after 400 years of Spanish effort to suppress native custom, and in spite of the fact that the Gregorian calendar now marks the Catholic holidays, which follow a separate cycle from the native calendar celebrations.
>
> LINCOLN 1942:103

Without dealing exhaustively with this extraordinary case of cultural continuity among the Ixil, we will consider the most important aspects of their calendar and its influence on the natives' lives. In the first place, we must take into account that the different units of time measurement used by the contemporary Ixil conform with similar measurements followed by the ancient Maya. Those units of time measurement are: a) the *ualyab* (*haab* in the ancient calendar) which is the count of the 365-day year; b) the *tzolkin* or 260-day cycle; c) the 360-day period called *tun* in former times; d) the 20-day month, and, e) the five complementary days forming the period called by the Ixil *o'ki,* which is the same as that known to pre-Columbian Maya as *uayeb* or *xma-kaba-kin.* The Ixil also recognize the four basic days with which a year may begin, the so-called "yearbearers"; today these are named *ij yab,* or "Mayors of the world." Contrasting with the Tzeltal Indians, the people of this area recall the names of the twenty days of the month. In reference to the names of the months, however, with the exception of one native priest who remembers them all, other people can only recall twelve or, at most, thirteen. The numbers used in the *tzolkin* count are always expressed in the Ixil language. This *tzolkin* count is computed by special priests, whereas the *haab* count is the concern of other wise men.

The numbers from one to thirteen, as well as the twenty days comprising the month, are considered to be living beings or deities to whom offerings and prayers must be rendered. The count of thirteen units of twenty days each constitutes the *tzolkin* or sacred year. Each date is expressed with a number followed by the name of the day. The sequence from one to thirteen, accompanied by the different days, is repeated indefinitely until the 260-day is reached. Then a new *tzolkin* begins. The count corresponding to the *haab* determines the day's position within the month. Within this system, to cite a complete date includes the same elements found in the pre-Columbian calendar. To wit: the number and day-name in the *tzolkin,* followed by the position of the date within its corresponding month.

The same occurred among the ancient Maya: as an example may be cited the well-known date of "4-*Ahau* 8-*Kumku,*" considered to be the initial point of their chronology. This date indicates that a day called *Ahau* occupied the fourth position within the thirteen-day period of the *tzolkin* and, at the same time, was the eighth of the twenty days comprising the month of *Kumku.* This would be analogous to saying "third Sunday, eighteenth of June"; in other words, the position of a day within our week of seven days and its position within our month. Naturally enough, to this date would have to be added the corresponding year as was done by the Maya within their system.

Burdens and fates of the days in the tzolkin

This period of 260 days constituted, fundamentally, a ceremonial and divinatory calendar, since by means of it propitious dates for traditional ceremonies were indicated through the calculations of the *Ah-kin,* or Priest of Time. Each day had its "burden," good or evil, affecting the destiny of human beings.

We will present here a list of the twenty days of the month used by the contemporary Ixil with the meaning of each day and its corresponding burdens. This list is based on the research of the already quoted J. Steward Lincoln.

Day Name	*Meaning and Corresponding Fate or "Burden"*
E	*Teeth.* Name of one of the "yearbearers." It is a day favorable to administer justice and to pray.
AJ	*Reed.* Appropriate for asking God to bless children. Firecrackers are exploded to beg protection for arms and shotguns. An auspicious day to pray for domestic fowl in general.

I'SH — *Ear of maize bears grains.* Day favorable for sheep, and for begging for the multiplication of animals. Day of the Lord of the goats.

TZIKIN — *Bird.* Favorable for chickens and for praying for money and other benefits.

A'MAK — *Sinner.* Favorable for the cornfields, for the "white cornfield."

NOJ — *Mayor.* One of the four possible "yearbearers" or "dominical day." Appropriate for praying for small domestic animals.

TIJASH — *Day of sacrifice.* Favorable for pigs, cows, and sacrificial animals. It is the day of the "Shepherd Foreman Cow-herd." (This is a deity frequently mentioned in prayers.)

KAUOK — *Guardian of the world.* Favorable for compensating for damage done by one's animals.

HUNAHPU — *Name of a god.* Saints' day and, also, of the Holy Virgin. Favorable for offering candles.

IMUSH — *Earth.* Favorable for praying in behalf of the home and family in order that the world or the earth not chastise them.

IK — *Wind.* One of the four possible "yearbearers." Lord of the wind. Day of the winds. Favorable for sowing corn.

AKBAL — *Night.* Bad. Day for doing evil to others. Sorcerer's day.

KATCH — *Net.* Bad. Day for harming others.

KAN — *Snake.* Favorable for requesting wealth.

KAMEL — *Death.* Favorable for the "yellow cornfield."

TCHE *Horse or deer.* One of the four possible "yearbearers." Very favorable.

KANIL *Ear of maize.* Day of the holy cornfield, favorable for all kinds of seeds and crops.

TCHO *Paid.* Day for mitigating conflicts with others, used especially by sorcerers. Day for recognizing sins and faults.

TCHII *Dog.* Bad. A day favored by sorcerers.

BATZ *Monkey.* Day of celebrations. Favorable to augment all things.

In this list, capital letters have been used to indicate those days which can become "yearbearers." As can be seen, each one is separated by five spaces from the other. Because the last five days of the year are not taken into account, and because 360 is a multiple of twenty (the days of the month), it is clear that if a year begins on a specific day, the following year will begin on a day five spaces from the previous one, and so on successively. In this way the "yearbearers" can only be four. The Ixil attribute the greatest meaning to these days and dedicate to them numerous ceremonies.

Furthermore, the natives always show deep interest in having the calendar diviners and priests solve an enigma or special problem. The questions most frequently heard refer to: the identity of the day, the fate of the same, the *nagual* (animal companion) which corresponds to a child born on that date; the diagnosis of an illness; seasons for sowing, harvesting, and performing other tasks in the cornfield; days favorable for an ill person to confess to his family; whether a trip will be lucky or not, as well as other problems. Sometimes the specialist uses some red beans or fragments of rock crystal for his divination. Depending on the attributes of each day, the specialist may also disclose the particular fortune corresponding to each person; thus, someone born on the day *Aj* (reed) will be poor; those born on the day *Tchii* (dog) will be lustful and sinful, and so on. In a list of days collected by Harry and Lucille McArthur (1965:33–38) in Aguacatan (near the Ixil) will be found a more detailed description of the fortune associated with each day and its corresponding ceremonies.

The cult of the yearbearer

We have seen that the name of "yearbearers" is given to those days with which the *haab* or 365-day year, can begin. As was explained, there are only four, and among the Ixil they are called *E, Noj, Ik,* and *Tche.* When the year commences there is a grand celebration in honor of its "bearer." Because of the structure of the calendar the same yearbearer day name must reappear 260 days later. Then another solemn feast will take place. The latter are the two most important yearly celebrations. The ceremonies are begun after sunset on the day preceding the *haab* date, since the natives consider this the first hour of the day.

In Nebaj, the principal Ixil town, many of the yearbearer ceremonies occur in front of the crosses distributed toward the four cardinal points or the "four corners of the world." At each of these points there is a cross placed atop an archaeological mound. These four sites are ordered in a rigid hierarchy, according to the following list:

1. *Tii Cajal:* "In front of the mound where blood flows."
2. *Tii Cuishal:* "In front of the mound where they danced."
3. *Cuchulchim:* "Where the kings assembled in ancient times."
4. *Chaxbatz:* "The green monkeys."

It is unfortunate that Steward Lincoln does not indicate the cardinal direction corresponding to each cross and mound. In front of these mounds, the ceremonies corresponding to the yearbearers are held by turn. Thus, to cite a case observed in 1939, the first day of the year corresponded to the day *E,* for which the ceremonies were celebrated in front of the most important mound, *Tii Cajal.* The other three potential yearbearers, inactive in that year, had their respective ceremonies in front of the mounds as follows: *Noh* in front of the cross of *Cuishal; Ik* before the cross of *Cuchulchim,* and *Tche* in front of the cross of *Chaxbatz.* The following year they jumped one place, and *Noh* came to occupy the site of *Tii Cajal, Ik* that of *Cuishal,* and *Tche* that of *Cuchulchim,* leaving the last place to the bearer *E.* As is obvious, in the course of four years, the rotation is complete. These data recall the proceeding followed by the Yucatec Maya described by Landa (Tozzer 1941:135–150) when referring to ceremonies that took place on the days of *uayeb* at the end of the year.

There also exists among the Ixil, although with little vitality, memory of the fifty-two–year cycle intimately connected with the concept of the yearbearers. It must be remembered that each yearbearer advances one place each year. As the yearbearers are four, four periods of thirteen years go by accompanied by as many numerals, thus giving origin to the fifty-two–year cycle well known in Central America.

The data presented above have shown the most relevant aspects of the aboriginal calendar in use today among various modern Maya peoples. We trust that our discussion has led the reader to grasp the degree to which these people have clung to the ancient concepts guiding them in their quest for the meaning of time. The ancient wisdom of the Maya was not totally destroyed: the revelations of time are still tied to the destiny of man.

BIBLIOGRAPHICAL REFERENCES

Anders, Ferdinand:

1963 *Das Pantheon der Maya.* Akademische Druck und Verlaganstalt, Graz.

Barrera Vásquez, Alfredo:

1941 "Sobre la identificación de algunos nombres de signos del calendario maya," *Los mayas antiguos,* pp. 79–86, México.

1948 *El libro de los libros de Chilam Balam.* Fondo de Cultura Económica, México.

1965 *El libro de los cantares de Dzitbalché.* Instituto Nacional de Antropología e Historia, México.

Barrera Vásquez, A., and Morley, S. G.:

1949 *The Maya Chronicles.* Carnegie Institution of Washington, Publication 585, Washington, D.C.

Barthel, T. S.:

1952 "Der Morgensternkult in der Darstellungen der Dresdener Mayahandschrift," *Ethnos,* vol. 17, pp. 73–112, Stockholm.

Becerra, Marcos E.:

1933 "El antiguo calendario Chiapaneco," *Universidad de México* (1a. serie), t. v, nums. 29–30, pp. 291–364, México.

Berlin, Heinrich:

1959 "Glifos nominales en el sarcófago de Palenque," *Humanidades,* vol. 2, num. 10, pp. 1–18, Guatemala.

Beyer, Hermann:
1930 "The Analysis of the Maya Hieroglyphs," *Internat. Archiv für Ethnologie,* 31, pp. 1–20, Leiden.
1932 "The Stylistic History of the Maya Hieroglyphs," *Middle American Research Series.* Publication 4, pp. 71–102, New Orleans.

Blom, Frans:
1954 "Ossuaries, Cremation and Secondary Burials among the Maya of Chiapas, Mexico." *Journal de la Société des Americanistes.* Musée de l'Homme, Paris.
1956 "Vida precortesiana del indio chiapaneco de hoy," *Estudios antropológicos en homenaje del Dr. Manuel Gamio,* pp. 277–285, México.

Bricker, Victoria R.:
1966 "El hombre, la carga y el camino: antiguos conceptos mayas sobre tiempo y espacio y el sistema zinacanteco de cargos," *Los zinacantecos,* Evon Z. Vogt (editor), Instituto Nacional Indigenista, México.

Caso, Alfonso:
1946 "Calendario y escritura de las antiguas culturas de Monte Albán," *Obras completas de Miguel O. de Mendizábal,* Talleres Gráficos de la Nación, vol. I, pp. 113–145, México.
1958 "El calendario mexicano," *Memorias de la Academia Mexicana de la Historia,* t. XVII, num. 1, pp. 41–96, México.
1965 "Zapotec Writing and Calendar," *Handbook of Middle American Studies,* vol. 3, pp. 931–947, Univ. of Texas Press, Austin.

Chilam Balam, Books of:
(See: Barrera Vásquez, Makeson, M. W., and Roys, R. L.)

Codex Dresden:

1880 *Die Maya-Handschrift der Königlichen Bibliothek zur Dresden;* herausgegeben von Prof. Dr. E. Förstemmann, Leipzig (2a. ed., 1892).

Codex Madrid (Cortesiano section):

1892 *Códice maya denominado Cortesiano* que se conserva en el Museo Arqueológico Nacional (Madrid). Reproducción fotocromolitográfica ordenada en la misma forma que el original. Edited by D. Juan de Dios de la Rada y Delgado and D. Jerónimo López de Ayala y del Hierro, Madrid.

Codex Madrid (Troano section):

1869–1870 *Manuscript Troano.* Études sur le systéme graphique et la langue des Mayas. Ed. C. E. Brasseur de Bourbourg, 2 vols., Paris.

Codex Paris:

1887 *Codex Peresianus,* manuscrit hiératique des anciens Indians de l'Amérique Centrale conservé a la Bibliothéque Nationale de Paris, avec une introduction par Léon de Rosny. Publié en coleurs, Paris.

Códice Pérez:

1949 *Códice Pérez,* textos mayas con versión castellana de E. Solís Alcalá, Mérida.

Coe, Michael D.:

1957 "Cycle 7 Monuments in Middle America: A Reconsideration," *American Anthropologist,* vol. 59, 4, pp. 597–611.

1965a "The Olmec Style and Its Distributions," *Handbook of Middle American Studies,* vol. 3, pp. 739–775, Univ. of Texas Press, Austin.

1965b "A Model of Ancient Community Structure in the Maya Lowlands," *Southwestern Journal of Anthropology,* vol. 21, 2, pp. 97–114, Univ. of New Mexico, Albuquerque.

Cogolludo, Diego López de:
1954 *Historia de Yucatán,* Comisión de Historia, Campeche.

Córdova, Fray Juan de:
1942 *Vocabulario en lengua zapoteca,* México.

Diccionario de Motul:
1929 Maya-español, atribuido a fray Antonio de Ciudad Real y Arte de la lengua . . . ed. J. Martínez Hernández, Mérida.

Dieseldorf, E. P.:
1926–1933 *Kunst und Religion der Mayavölker im alten und heutigen Mittelamerika,* 3 vols., Berlin.

Durán, Fray Diego de:
1964 *The Aztecs: the History of the Indies of New Spain,* translated, with notes by Dorys Heiden and Fernando Horcasitas, Orion Press, New York.

Förstemmann, E. W.:
1902 *Commentar zur Madrid Handschrift,* Danzig.
1903 *Commentar zur Pariser Handschrift,* Danzig.
1906 *Commentary on the Maya Manuscript in the Royal Public Library of Dresden,* Papers, Peabody Museum, vol. 4, 2, Cambridge, Mass.

Fuente, Beatriz de la.
1964 *La escultura de Palenque,* Instituto de Investigaciones Estéticas, Universidad Nacional, México.

Gillin, John:
1958 *San Luis Jilotepec,* Seminario de Integración Social de Guatemala, pub. num. 7, Guatemala.

Girard, Rafael:
1962 *Los mayas eternos,* Antigua Librería Robredo, México.

Goodman, J. T.:
1889–1902 *The Archaic Maya Inscriptions,* appendix Maudsley, A. P., *Archaeology, Biologia Centrali-Americana,* 5 vols., London.

Gordon, G. B.:

1902 "On the Use of Zero and Twenty in the Maya Time System," *American Anthropologist,* 4, pp. 237–275.

Goubaud Carrera, Antonio:

1935 "El *Guajxaquib Batz,* ceremonia calendárica indígena," *Anales de la Sociedad de Geografía e Historia de Guatemala,* tomo XII, num. 1, pp. 39–50.

Guiteras Holmes, C.:

1947 "Clanes y sistemas de parentesco de Cancuc, México," a reprint of *Acta Americana,* vol. v, num. 1–2.

1961 *Perils of the Soul,* The Free Press of Glencoe.

1965 *Los peligros del alma: visión del mundo de un tzotzil,* Fondo de Cultura Económica, México.

Holland, William R.:

1963 *Medicina maya en los Altos de Chiapas,* Instituto Nacional Indigenista, México.

Kaufman, Terrence S.:

"Materiales lingüísticos para el estudio de las relaciones internas de la familia de idiomas mayanos," *Desarrollo cultural de los mayas,* Universidad Nacional, Seminario de Cultura Maya, pp. 81–136, México.

Kelley, David H.:

1965 "The Birth of the Gods of Palenque," *Estudios de cultura maya,* Universidad Nacional, Seminario de Cultura Maya, vol. V, pp. 93–134, México.

Krickeberg, Walter:

1961 *Las antiguas culturas mexicanas,* Fondo de Cultura Económica, México.

La Farge, Oliver:

1947 *Santa Eulalia: The Religion of a Cuchumatan Indian Town,* Univ. of Chicago Press, Chicago.

1962 "Maya Ethnology: The Sequence of Cultures." In *The Maya and Their Neighbors.* D. Appleton-Century Co., New York.

La Farge, Oliver, and Byers, Douglas S.:

1931 *The Year Bearer's People,* Tulane University of Louisiana, Middle American Research Series, pub. num. 3, New Orleans.

Landa, Fray Diego de:

1938 *Relación de las cosas de Yucatán,* edited by A. Pérez Martínez, Editorial Pedro Robredo, México.

1941 *Landa's Relación de las cosas de Yucatán,* a translation, edited with notes by A. M. Tozzer, Papers, Peabody Museum, vol. 18, Cambridge, Mass.

León-Portilla, Miguel:

1963 *Aztec Thought and Culture,* Univ. of Oklahoma Press, Norman.

Lincoln, J. Steward:

1942 *The Maya Calendar of the Ixil of Guatemala,* Carnegie Institution of Washington Publications, Contribution 528.

Lizardi Ramos, César:

1936 *Recurrencias de las fechas mayas,* México.

1962 "El cero maya y su función," *Estudios de cultura maya,* vol. II, Universidad Nacional, Seminario de Cultura Maya, pp. 343–353, México.

López de Cogolludo, Diego:

(See: Cogolludo, Diego López de).

Makemson, M. W.:

1951 *The Book of the Jaguar Priest,* a translation of the book of *Chilam Balam* of Tizimin, H. Schuman, New York.

Maudsley, A. P.:

1889–1902 *Archaeology, Biologia Centrali-Americana,* 5 vols., London.

McArthur, Harry and Lucille:

1965 *Notas sobre el calendario ceremonial de Aguacatán,* Huehuetenango, Instituto Lingüístico de Verano, Folklore de Guatemala, num. 2, Guatemala, pp. 23–38.

McQuown, Norman A.:

1956 "The Classification of the Mayan Languages," *International Journal of American Linguistics,* vol. 22, pp. 191–196.

1964 "Los orígenes y la diferenciación de los mayas según se infiere del estudio comparativo de las lenguas mayanas," *Desarrollo cultural de los mayas,* Universidad Nacional, Seminario de Cultura Maya, pp. 49–80, México.

Miles, S. W.:

1957 *The Sixteenth-Century-Pokom-Maya: A Documentary Analysis of Social Structure and Archaeological Setting,* Transactions, part 4. Independence Square, Philadelphia.

Molina, Marta F. de:

1965 *La escultura arquitectónica de Uxmal,* Instituto de Investigaciones Estéticas, Universidad Nacional, México.

Morley, Silvanus G.:

1911 "The Historical Value of the Books of Chilam Balam," *American Journal of Archaeology* (2nd series), num. 15, Norwood.

1947 *The Ancient Maya,* Stanford Univ. Press, California.

Núñez de la Vega, F.:

1702 *Constituciones diocesanas del Obispado de Chiapas,* Roma.

Oakes, Maud.:

1951 *The Two Crosses of Todos Santos,* Bollingen Series XXVII, Pantheon Books, New York.

Pío Pérez, Juan:

1937 "Antigua cronología yucateca," Appendix in *Historia antigua de Yucatán,* by Crescencio Carrillo y Ancona, Mérida.

Popol Vuh:

(See: Recinos, A., and Schultze-Jena, L.)

Recinos, Adrián:

1947 *El Popol Vuh: las antiguas historias del Quiché,* Fondo de Cultura Económica, México.

1961 *Popol Vuh, the Sacred Book of Ancient Quiché Maya,* Univ. of Oklahoma Press, Norman.

Redfield, Robert, and Villa Rojas, Alfonso:

1962 *Chan Kom: A Maya Village,* Univ. of Chicago Press.

Relaciones de Yucatán:

1898–1900 in *Colección de documentos inéditos relativos al descubrimiento, conquista y organización de las antiguas posesiones españolas de ultramar,* 2nd. Series, vols. 11 and 13, Madrid.

Remesal, Antonio de:

1932 *Historia de la provincia de San Vicente de Chiapas y Guatemala,* 2 vols., Sociedad de Geografía e Historia, Guatemala.

Roys, Ralph L.:

1933 *The Book of Chilam Balam of Chumayel.* Carnegie Institution of Washington, pub. 438, Washington, D.C.

1943 *The Indian Background of Colonial Yucatán,* Carnegie Institution of Washington, pub. 548, Washington, D.C.

1946 *The Book of Chilam Balam of Ixil,* Carnegie Institution of Washington. Notes on Middle American Archaeology and Ethnology, 75, Cambridge, Mass.

1965 *Ritual of the Bacabs,* Univ. of Oklahoma Press, Norman.

Ruz Lhuillier, Alberto:

1954 "Exploraciones en Palenque: 1952," *Anales del Instituto Nacional de Antropología e Historia,* t. VI, pp. 79–112, México.

1963 *La civilización de los antiguos mayas,* Instituto Nacional de Antropología e Historia, México.

Sánchez de Aguilar, Pedro:

1937 *Informe contra Idolorum Cultores,* 3rd. edition, Mérida.

Satterthwaite, L.:

1947 *Concepts and Structures of Maya Calendrical Arithmetics,* Philadelphia.

Schellhas, P.:

1904 *Representation of Deities of the Maya Manuscripts,* Papers of the Peabody Museum, vol. 4, num. 1, Cambridge, Mass.

Schultze-Jena, Leonhard:

1944 *Das Popol Vuh, Das heilige Buch der Quiche Indianer von Guatemala,* Quellenwerke zur alten Geschichte Amerikas, vol. II, Stuttgart.

Seler, Eduard:

1902–1923 *Gesammele Abhandlugen zur Amerikanischen Sprachund Altertumskunde,* 5 vols., Ascher und Co. and Behrend und Co., Berlin.

Siegel, Morris:

1941 "Religion in Western Guatemala: A Product of Acculturation," *American Anthropologist,* vol. 43, num. 1, pp. 62–76.

Sodi, Demetrio:

1964 *La literatura de los mayas,* Joaquín Mortiz, Instituto Indigenista Interamericano, México.

Solís Alcalá, Hermilo:

1949 *Códice Pérez. Traducción libre del maya al castellano,* Mérida, Yuc., México.

Soustelle, Jacques:
1937 *La Culture Matérielle des Indiens Lacandons,* Paris.
Spinden, Herbert J.:
1957 *Maya Art and Civilization,* 2nd ed., The Falcon's Wing, New York.
Swadesh, Mauricio:
1961 "Interrelaciones de las lenguas mayas," *Anales del Instituto Nacional de Antropología e Historia,* México.
Teeple, John E.:
1930 *Maya Astronomy,* Carnegie Institution of Washington, pub. 403, Contribution 2, Washington, D.C.
Thompson, J. Eric S.:
1930 *Ethnology of the Mayas of Southern and Central British Honduras,* Field Museum of Natural History, Anthropol. Ser., vol. 17, num. 2, Chicago.
1934 *Sky Bearers, Colors and Directions in Maya and Mexican Religion,* Carnegie Institution of Washington, Publication 436, contrib. 10, Washington, D.C.
1939a "Las llamadas fachadas de Quetzalcóatl," *Memorias del XXVII Congreso Internacional de Americanistas,* pp. 341–400, México.
1939b *The Moon Goddess in Middle America,* With Notes on Related Deities, Contributions to American Anthropology and History, num. 29, Carnegie Institution of Washington.
1941 *Maya Arithmetic,* Carnegie Institution of Washington, Publication 36, Washington, D.C.
1950 *Maya Hieroglyphic Writing,* Carnegie Institution of Washington, Washington, D.C.
1952 "The Character of the Maya," *Proceedings of the XXXth International Congress of Americanists,* Royal Anthropological Institute, London.
1954 *The Rise and Fall of Maya Civilization,* Univ. of Oklahoma Press, Norman.

1960 *Maya Hieroglyphic Writing: An Introduction,* Univ. of Oklahoma Press, Norman.

1962 *A Catalog of Maya Hieroglyphs,* Univ. of Oklahoma Press, Norman.

Tozzer, Alfred M.:

1907 *A Comparative Study of the Mayas and the Lacandones,* The Macmillan Company, New York.

1921 *A Maya Grammar with Bibliography and Appraisement of the Works Noted,* Papers, Peabody Museum, Harvard University, vol. 9, Cambridge, Mass.

1941 *Landa's relación de las cosas de Yucatán,* a translation, edited with notes, Papers, Peabody Museum, Harvard University, vol. 18, Cambridge, Mass.

Tozzer, A. M., and Allen, G. M.:

1910 *Animal Figures in the Maya Codices,* Papers, Peabody Museum, Harvard University, vol. 4, num. 3, Cambridge, Mass.

Valladares, León A.:

1957 *El hombre y el maíz; etnografía y etnopsicología de Colotenango,* Guatemala, C. A.

Villa Rojas, Alfonso:

1941 "Dioses y espíritus paganos de los mayas de Quintana Roo," *Los mayas antiguos,* Colegio de México, México.

1945 *The Maya of East Central Quintana Roo,* Carnegie Institution of Washington, Publication 559, Washington, D.C.

1946 *Notas sobre la etnografía de los indios tzeltales de Oxchuc, Chiapas,* microfilm coll. of mss. on Middle American Cultural Anthropology, num. 7, Univ. of Chicago.

1947 "Kinship and Nagualism in a Tzeltal Community, Southeastern Mexico," *American Anthropologist,* vol. 49, num. 4, part 1, pp. 578–587.

1956 "Notas introductorias sobre la condición cultural de los mijes," Introduction to *Cuentos Mixes* of

Walter Miller, Instituto Nacional Indigenista, México.

1961 "Notas sobre la tenencia de la tierra entre los mayas de la antigüedad," *Estudios de cultura maya,* vol. I, pp. 21–46, Universidad Nacional, Seminario de Cultura Maya, México.

Vogt, Evon Z.:

1964 "The Genetic Model and Maya Cultural Development," *Desarrollo cultural de los mayas,* ed. by E. Z. Vogt and A. Ruz, Seminario de Cultura Maya, Universidad Nacional, pp. 9–48, México.

1966 "Conceptos de los antiguos mayas en la religión Zinacanteca contemporánea," chapter III of *Los Zinacantecas,* Instituto Nacional Indigenista, México.

1969 *Zinacantan, a Maya Community in the Highlands of Chiapas,* Harvard University Press, Cambridge, Mass.

Wagley, Charles:

1957 *Santiago Chimaltenango, Guatemala,* Seminario de Integración Social Guatemalteca, Guatemala.

Weitlaner, Roberto J.:

1939 "Los Chinantecos," *Revista Mexicana de Estudios Antropológicos,* t. III, n. 3, Sociedad Mexicana de Antropología, México.

Ximénez, Francisco:

1929–1931 *Historia de la provincia de San Vicente de Chiapas y Guatemala de la Orden de Predicadores,* Biblioteca Goathemala, vols. 1–3, Guatemala.

Zimmermann, Günter:

1956 *Die Hieroglyphen der Maya-Handschriften,* Universitat Hamburg Abhandlungen aus dem Gebiet der Auslandkunde, vol. LXVII, Reihe B, Hamburg.

INDEX